This book is dedicated to the great teachers
in my life, especially Donald J. Borror,
Theodore H. Hubbell, Ernest M. R. Lamkey,
Blanche McAvoy, Donald R. Meyer, Edward S. Thomas, John N. Wolfe, and, most
of all, my parents, Archie Dale Alexander
and Katharine Elizabeth Heath.

The Jessie and John Danz Lectures

THE JESSIE AND JOHN DANZ LECTURES

DARWINISM
AND
HUMAN AFFAIRS

RICHARD D. ALEXANDER

University of Washington Press
Seattle and London

Library of Congress Cataloging in Publication Data

Alexander, Richard D
  Darwinism and human affairs.

  (The Jessie and John Danz lectures)
  Bibliography: p.
  1. Social evolution. 2. Evolution.
3. Natural selection. I. Title. II Series.
HM 106.A44   301'.0424   78–65829
ISBN 0–295–95641–0 (cloth); 0–295–95901–0 (paper)

The Jessie and John Danz Lectures

In October 1961, Mr. John Danz, a Seattle pioneer, and his wife, Jessie Danz, made a substantial gift to the University of Washington to establish a perpetual fund to provide income to be used to bring to the University of Washington each year "distinguished scholars of national and international reputation who have concerned themselves with the impact of science and philosophy on man's perception of a rational universe." The fund established by Mr. and Mrs. Danz is now known as the Jessie and John Danz Fund, and the scholars brought to the University under its provisions are known as Jessie and John Danz Lecturers or Professors.

Mr. Danz wisely left to the Board of Regents of the University of Washington the identification of the special fields in science, philosophy, and other disciplines in which lectureships may be established. His major concern and interest were that the fund would enable the University of Washington to bring to the campus some of the truly great scholars and thinkers of the world.

Mr. Danz authorized the Regents to expend a portion of the income from the fund to purchase special collections of books, documents, and other scholarly materials needed to reinforce the effectiveness of the extraordinary lectureships and professorships. The terms of the gift also provided for the publication and dissemination, when this seems appropriate, of the lectures given by the Jessie and John Danz Lecturers.

Through this book, therefore, another Jessie and John Danz Lecturer speaks to the people and scholars of the world, as he has spoken to his audiences at the University of Washington and in the Pacific Northwest community.

It falls to a journalist reviewing the books of our day to treat the dreadful almost as though it were commonplace. The books I review, week upon week, report the destruction of the land or the air; they detail the perversion of justice; they reveal national stupidities. None of them—not one—has saddened me and shamed me as this book has. Because the experience of reading it has made me realize for once and all that we really don't know who we are, or where we came from, or what we have done, or why.

Geoffrey Wolff, *Newsweek* (from a review
of *Bury My Heart at Wounded Knee*)

Preface

My purpose, in the lectures from which this book was developed, was to review and extend our understanding of the relationship between the process of organic evolution and the structure, variations, and significance of human behavior. I wished to show that a thorough understanding of the evolutionary processes by which life has formed and diversified is a vital part of the everyday knowledge of ordinary human beings in a society such as ours—that is to say, those who vote in a democracy, those who help guide social, cultural, and technological change, those who influence the lives of other human beings, and those who reflect upon their own existences and strive consciously to become the masters of their own fates. I emphasized what I see as the philosophical significance of evolution because it is defensible as the only theoretical base from which to undertake a truly comprehensive analysis of human activities and tendencies. I also sought to diminish the reasons for the existing hostilities and strife that derive, as I see it, from miscommunications and unnecessary disagreements between biological and social scientists, and between biologists and humanists, in regard to evolution and its ideological and ethical meanings.

Within the past two decades, refinements of evolutionary

theory have created a major revolution in biology and produced an entirely new view of human nature. I think it is worthwhile to state briefly what happened to cause this revolution, and to explain the reasons for its philosophical impact.

The notion of genes as heritable factors or units that recombine during sexual reproduction and, together with the environment, shape the organism's behavioral, physiological, and structural traits dates almost—but not quite—to Darwin, through the work of the Austrian monk Gregor Mendel. As a formal concept integrated into biology, this view of genes is only about as old as the twentieth century, although still a half century older than the knowledge that the genetic material is comprised of deoxyribonucleic acid molecules. The evolutionist's view of the genes as recombining units is also a half century older than the efforts of molecular biologists to begin defining genes directly from knowledge of their physiological activities or effects.

Not until the 1960s did social biologists realize how important it is for them to understand that what natural selection has apparently been maximizing is the *survival by reproduction* of the genes, as they have been defined by evolutionists, and that this includes effects on the copies of genes, even when those copies are located in other individuals. In other words, selection has not operated so as to promote directly either the long-term survival of individuals or the well-being and survival of populations or species at the expense of individuals. When such effects occur—whether as a result of social interactions or for other reasons—they are either incidental consequences of the evolutionary process or they arise out of such consequences.

In one stroke this realization provided the means for solving numerous long-standing biological problems involving sexuality, senescence, sex ratios, male-female interactions, parental care, and nepotism. Perhaps most importantly, it resolved the ancient philosophical paradox whether humans are really selfish individualists or group altruists, and provided, I believe, the first simple, general theory of human nature with a high likelihood of widespread acceptance. The answer to the age-old riddle that even ordinary humans ask themselves appears to be that we are selfish individualists in the sense and to the extent that this maximizes the survival by reproduction of the genes residing in our own bodies, and we are group altruists in the sense and to the extent that this maximizes the survival by reproduction of the copies of our genes residing in the bodies of others—that is, in the bodies of our genetic relatives, both descendant and nondescendant. At least this is what we have evidently evolved to be, in the usual environments of history, notwithstanding (1) the difficulties of determining what it means that our environments have surely been altered by culture and technology far too rapidly for our evolved phenotypes (or developmental potentials) to "track" them adaptively and (2) our inability to know what the new knowledge about our nature and our history may enable us to *become.*

Human behavior is exceedingly complex and diverse. It follows that if an appropriate and reasonable theory of human nature can be stated simply, it surely will be exceedingly difficult to apply. This is obviously true with the new theory from evolutionary biology. So long as human behavior and the capacity to generate, absorb, use, and maintain culture were seen as evolved to contribute to the well-being of the social group, population, or species as a whole, conflicts of interest were relatively simple: by definition they should exist only as accidents or mistakes below the level at which function was being ascribed. The new theory, however, drives the problem of understanding conflicts of interest right to the level of the gene. This means that, on evolutionary grounds, individual humans are expected to behave as though they have individualized sets of interests because they have individualized sets of genes. The degrees of difference in their interests—and therefore their tendencies to cooperate and to be altruistic toward one another—can now be quantified. We expect them to correlate, in the usual environments of history, with degrees of genetic relatedness, or the general likelihood that any gene is shared between them. The major *seeming* exception is that means have developed for enforcing reciprocity, or payment in systems of exchange, causing such systems, operative even among nonrelatives, to become a major aspect of human sociality. Of course, such systems do not really constitute exceptions to evolution, since systems of reciprocity evolve too, but they lead to outcomes predictably different from those involved in nepotism.

By the new evolutionary theory of human nature, then, the analysis of conflicts of interest simultaneously takes on a new aspect, because of our knowledge of the genetic asymmetries that underlie such conflicts, and becomes recognizable as both a stupendously difficult task and the only apparent way to understand in depth what human sociality and culture are all about.

Darwin did not know about genes so he could not fully resolve the problem of human tendencies in different circumstances to be either altruistic or selfish; nor were the genes, in the sense discussed here, a part of Freud's understanding, or of that of any other major philosopher or student of human nature before this decade. Genes are not yet a part of the everyday consciousness of a significant fraction of even the most thoughtful and best educated people of the world. It is astounding—even bizarre—to realize that through the whole rise of human consciousness and self-analysis, the tiny objects that underlie it all remained outside the reach of our senses until the last few decades, and the true nature of their mission was still significantly blurred right into the current decade. This necessarily means that we are evolved to carry out our everyday social transactions without knowing exactly why we do what we do. It is a matter of deep intrigue to ponder the consequences of bringing the evolutionary reasons for our existence and our nature fully into our consciousness, and of dwelling upon them at length.

For several reasons I regard it as especially relevant that biologists take up the problem of relating human attributes to evolutionary history. First, of course, is the unknowably immense potential for self-understanding to further the interests of humanity in general—indeed, even literally to enable its survival. Second, even if we humans find it difficult to view ourselves dispassionately, we know more about our own behavior than about that of any other species. Taken together with the fact that we display many unique behaviors, this means that the examination of human behavior has the potential of feeding back directly into biology insights, theories, and knowledge not as easily gained from studying any nonhuman species.

The third reason requires explanation. Anthropologists and evolutionary biologists proceed in parallel fashions. First, they study and compare variations (usually among cultures and species, respectively) and their probable significance or adaptive value. Then they search for connections between the adaptive significance and the immediate (physiological, ontogenetic) mechanisms responsible for the variations. The two kinds of investigators differ in that whenever they are unable to resolve the question of connections between adaptive significance and immediate mechanisms for particular behaviors (cultural patterns), social scientists may often assume that no such connections exist while biologists usually continue to assume that they do. It is not uncommon for nonscientists also to assume that human behavior is independent of evolutionary history. This assumption, by whomever it is made, provides an excuse for diminishing the significance of evolution as a guiding principle in understanding not only humans but life in general. It is in the interests of biologists—and I believe of all humans—that any such tendencies to reduce the central significance of organic evolution are continually forced to be grounded on knowledge and not ignorance.

For these reasons I regard it as especially useful for biologists to contribute to the analysis of human behavior on all reasonable fronts, especially when massive theoretical revolutions occur within their science; and that is certainly true at this moment in history.

Much of this book is derived from the approximately twenty papers on behavior and evolution that I have published since 1969. Because none of these papers is widely available to nonprofessionals I have sometimes used long passages almost verbatim. But there are also many changes and significant additions. In the first chapter I review the history and structure of evolutionary theory, as I see it, developing a step-by-step argument toward the conclusion that to understand our humanness we must concentrate primarily on the generation-to-generation process of evolutionary adaptation, and that we must know how to predict the long-term cumulative consequences of adaptive genetic change. This chapter is general, in that it considers the theory and process of evolution as a whole. It is also specialized, for it does not attempt to deal comprehensively with the vast literature and evidence surrounding evolution. Rather, I have utilized a narrowly selected set of examples to engage deliberately what I see as the two principal classes of critics of evolution remaining today: philosophers and other academic and intellectual nonbiologists, many of whom do not regard evolution as a truly scientific and testable theory; and creationists and others, chiefly of one or another religious inclination, who continue to suggest (and believe) that evolution is a pernicious evil, or at least an unpleasant and erroneous idea.

In the second chapter I draw the connections between the process of organic evolution and the existence and modification of culture, arguing that cultural and genetic change have not been independent, or opposed to one another. I argue that culture is, like phenotypes among all organisms, a kind of biological adaptation through plasticity that is simply more elaborate and has a greater potential for intergenerational heritability than other kinds of phenotypic plasticity. In the third chapter I show that variations in cultural patterns, like phenotypic variations in general, are interpretable as outcomes of the reproductive strivings of competing and cooperating individuals who live in different circumstances. In the final chapter I explain my conviction that evolutionary biology is our best hope for knowing how to achieve whatever social or ethical goals we may set for ourselves. I also argue that it contains no simple or static answers to the pressing questions of how, in any era, place, or society, to identify what those goals ought to be. In short, this is a book about human natural history—the "was" and the "is" of our social behavior—and it says almost nothing about "what ought to be."

Acknowledgments

I am grateful to John S. Edwards and the Jessie and John Danz Lectureship Committee of the University of Washington for providing the stimulus and the opportunity to develop this book, by inviting me to lecture in Seattle in November and December of 1977.

I also appreciate the stimulative effects of all my graduate students and colleagues, and the constructive tolerance and freedom of inquiry afforded me in the University of Michigan's Museum of Zoology during the approximately ten years in which the effort summarized here was developed.

For reading and criticizing part or all of the manuscript I thank David Cowan, Steven Frank, Elizabeth McLeary, Joan Miller, M. F. Ashley Montagu, Joel Peck, Elizabeth Rockwell, Donna Silverman, Kenneth Warheit, Mary Jane West Eberhard, the members of my 1978 and 1979 seminars on evolution and human behavior, and, especially, D. Caldwell Hahn, William D. Hamilton, William G. Irons, Bobbi S. Low, Ann E. Pace, Paul W. Sherman, and John Strate. Napoleon A. Chagnon, Mildred Dickemann, Mark V. Flinn, Kent V. Flannery, Richard D. Howard, Katharine M. Noonan, Elizabeth Rockwell, Bruce Wallace, and Henry T. Wright assisted with special problems. Mary Snider patiently and skillfully typed and retyped the manuscript. Mark J. Orsen prepared the illustrations. Joan Miller was superbly helpful with library work and proofreading. Theodore H. Hubbell spent many hours over coffee listening to my problems and helping solve them. My wife, Lorraine Kearnes Alexander, as always, kept her sympathetic and critical ear turned in my direction.

For permission to use passages from my previously published works I thank the Royal Society of Victoria, Melbourne; the Philosophy of Science Association; the Philadelphia Academy of Natural Sciences; the Hastings Institute of Society, Ethics, and the Life Sciences, and the National Endowment for the Humanities; Annual Reviews, Incorporated; *Behavioral Science; The American Biology Teacher;* the Duxbury Press; and the Academic Press.

I thank the Macmillan Company for permission to use passages from George Peter Murdock's *Social Structure;* E. S. Burch, Jr., and the West Publishing Company for permission to quote from *Eskimo Kinsmen;* Professor Vern Carroll and the University of Hawaii Press for permission to quote from *Adoption in Oceania;* the Harvard University Press for permission to quote from E. A. Hoebel's *The Law of Primitive Man;* Charles E. Silberman and Random House for permission to quote from *Criminal Law, Criminal Justice;* Professor Kent V. Flannery and Annual Reviews, Incorporated, for permission to quote from "The Evolution of Civilizations"; Katharine M. Noonan and Gerald Borgia for permission to use passages from jointly authored papers; and the Grove Press for permission to use lines from Jorge Luis Borges' "Poem Written in a Copy of Beowulf," from *Labyrinthes,* edited by Donald A. Yates and James E. Irby.

Contents

List of Illustrations

TABLES

DARWINISM AND HUMAN AFFAIRS

I

The Challenge of Darwinism

Scarcely anyone would deny that Darwin's theory of organic evolution by natural selection is one of the several great theories of all time. The words "Darwin" and "evolution" occur in the index of almost every book on philosophy or the history of science. On the other hand, the reference is frequently to a single page, where little more than a passing comment is given. Lengthy discussions of evolution are in fact still rare in the books that deal with the structure and significance of scientific theories, and they invariably terminate uncertainly or inconclusively.

Why should this be so? It is as if the historians and philosophers of science somehow know that evolution has to be one of the truly significant theories, yet they cannot quite tell what biologists are doing with the theory of evolution or how they are doing it.

Perhaps this is exactly the case. With the exception of divine creation, evolution is the only theory advanced to explain the existence and traits of life on earth. Yet life on earth is still the most complex phenomenon that we humans have encountered in the universe. For a theory to survive for more than one

3

hundred years as the sole proffered scientific explanation for the most complicated known phenomenon must be sufficient to qualify it as a great theory.

There are also reasons why evolution might be more difficult to understand than other great theories. Most theories that would be so regarded deal with questions about the nonliving physical universe: gravity, relativity, the nature of fundamental particles, or the behavior of gases. All of these theories combined purport to deal with a few hundred kinds of subatomic particles, about one hundred kinds of atoms or elements, and a few thousand kinds of fairly simple molecules that occur naturally outside living creatures. In contrast, the theory of evolution proposes to account for several million species of contemporary living organisms and at least a thousand times that many during the geological history of Earth. Some of these species contain billions of individuals, every one of which is unique. The number of different kinds and combinations of molecules involved is staggering.

Theories about nonliving phenomena may some day unlock secrets of the universe that are now beyond our wildest imagination, and they may deal with mechanisms that are in some sense more "basic" than those studied by biologists. Nevertheless, it remains true that the complexity of *organization of life,* which biological theories must explain, is nowhere approached in the yet observable nonliving world.

With a few exceptions like Freudian theory and theories of learning—themselves necessarily, in some sense, subsets of evolutionary theory—other major scientific theories do not threaten to invade or change our everyday existences, at least not in fashions that we are likely to regard as pernicious. They do not threaten to make our behavior predictable, to expose what we are actually doing in our social interactions, to infringe our concept of free will, or to influence the ways in which we think about right and wrong. Evolution does all of these things because it proposes to explain the explainers themselves. The difficulty of that proposition lies not only in the fact that some of the traits to be explained must be used in their own explanation, but also in that one trait of the explainers seems to be that they do not always wish to be explained—at least not too completely. In all likelihood, no theory about anything extrinsic in the universe will ever hold as much intrigue, or encounter as much resistance, as a theory about ourselves. It may be the ultimate irony of human existence that the more any such theory explains, the more difficult it will be to gain its widespread acceptance.

As a theory, then, evolution is unique in several regards. Small wonder that it has always been misunderstood by historians of science, philosophers of science, popularizers of science, and scientists themselves, including, I believe, the majority of biologists. Small wonder that it is almost the only great theory that has been maligned, ridiculed, and made into something analogous to pornography, not to be revealed to children while they are still young and impressionable.

In the last twenty years discoveries by biologists have made evolutionary theory more predictive than ever before. In this book I propose to show (1) how evolution can be used as an explanatory principle to account for human behavior, (2) how the process of evolution relates to the directions and rates of cultural change, and (3) how knowledge about the evolution of human behavior can be the most liberating of all possible advances in human understanding. I shall begin with the apparent variations in the meaning of the concept of organic evolution to different people.

Macro- and Microevolution

Darwin claimed to have come upon the idea of evolution by natural selection in the following way: in observing birds and other organisms from different populations, especially on the Galapagos Islands, he noticed that species on islands distant from one another tended to differ more from one another than did those on islands closer together. Species on islands distant from the mainland tended to differ more from mainland species than did those on islands close to the mainland. Apparently he realized that in some cases he could not even tell whether or not populations on neighboring islands represented different species. From these and other observations he gained the impression that species are not, after all, immutable, and that the variously different populations he was comparing represented the speciation process in several different stages. As early as 1844 he wrote to a friend, Joseph Hooker, that to admit to this revelation in those conservative times was "like confessing a murder."

Darwin thus envisioned that a single ancestral species might give rise to two daughter species over a very long period of time, and with that realization he evidently began to construct in his mind an ever-expanding phylogeny, or family tree, of living creatures. If we accept the sequence of thought suggested by Darwin's initial reference in 1837 in his *Evolutionary Notebook* to "Transmutation of Species," and the 1844 date of his letter to Hooker (see Lack, 1939), we can visualize the problem that he faced next: If there have been long-term patterns of change in the features of life on earth, what forces could account for them? What could cause the inexorable divergences that could lead to the multiplication of species and account for their diversity? Darwin's eventual answer was that since different forms, arising through the appearance of heritable changes, tend to reproduce at different rates, some would spread and persist while others would become extinct. He called this process, which is both theoretically sound and empirically verifiable, natural selection or "survival of the fittest."

Following the publication of Darwin's *On the Origin of Species* in 1859, biologists and paleontologists thoroughly examined the fossil record and compared living forms for evidence of the historical interconnection that Darwin presumed between living forms and their extinct ancestors. Although these scientists could never fill in completely the long-term phylogenetic pictures they were trying to reconstruct, ample evidence was soon uncovered that Darwin's theory was entirely reasonable. Paleontological data have since continued to support his view of the history of life in ever-increasing detail.

For most people the term evolution still refers largely or entirely to long-term changes in the *pattern* of life through geological time, as is evident from the fossil record and by the arrangement of similarities and differences among living organisms. Thus restricted, evolution seems not to be a predictive theory, because not enough is known of the environments of the distant past to reconstruct the forces of change operating then. We still do not know, in detail, the causes of even major prehistoric events like the extinction of the dinosaurs.

The concept of organic evolution, however, does not apply solely to long-term pattern changes, but also to the short-term changes resulting from the process of natural selection—to changes in gene frequencies from one generation to the next. Such changes are both directly and indirectly observable or measurable in contemporary organisms, and are sometimes termed "descent with modification." Biologists nowadays almost always mean both long- and short-term changes when they use the term evolution because they regard the long-term patterns and the current nature and diversity of life as simply the cumulative results of the short-term observable process. These long-term and short-term aspects have been labeled, by some, macroevolution and microevolution.

Recently, creationists and other antagonists to the concept of organic evolution have maintained that macro- and microevolution are entirely different things. They argue that because one cannot directly observe events like the formation of species or major organs, but must rely on circumstantial evidence, it is unscientific to suggest that long-term operation of the microevolutionary process leads to such events. Not many biologists take seriously the argument that circumstantial evidence is not scientific. But I call attention to a neglected challenge by Darwin (1859, p. 189) that anticipates this objection and leads to its dismissal. Darwin's statement far antedates methodologies proposed and made prominent by contemporary philosophers of science, some of whom have doubted the validity of evolution as a scientific theory because of what they see as the absence of suitable falsifying propositions or operations. Darwin said the following:

> If it could be demonstrated that any complex organ existed, which could not possibly have been formed by numerous, successive, slight modifications, my theory would absolutely break down.

This challenge is one of several which showed that Darwin was trying to postulate ways in which his theory could be falsified. We are safe in assuming that no such organ has been found since Darwin issued the challenge in 1859, and I am sure that most biologists would regard the search for an organ of this sort as futile. Nevertheless, the question Darwin raises has been tested, in a different way, simply by the accumulation of information about complex organs. Consider Darwin's challenge from another direction. If one should cross many forms (either within or between species) with variant versions of major organs and obtain large numbers of hybrids with a great variety of *slightly different forms* he obviously would have supported the argument that complex organs have been formed by numerous, successive, slight modifications, and, thus, the thesis that macroevolution is just extended microevolution. This "test" has inadvertently been carried out innumerable times by biologists and animal and plant breeders, and this kind of support of evolution has probably been more useful in biology than explicit efforts to *falsify* the universality of natural selection.

Similarly, if one were to cross different species or genera that do not interbreed in nature, and in the successive generations of hybrids obtain a great variety of slightly different kinds of individuals, he would also have connected micro- and macroevolution. Obviously, this test, as well, has been carried out many times by biologists. Both of these results are exactly what one would expect if the major groups of organisms arose through the preservation and accumulation of large numbers of very slight changes of the sort that we now know are owing to gene mutations; hence, they tend to falsify any other view of the history of life, such as divine creation, that is not similarly based (see also Alexander, 1978a).

Another challenge to the idea that macroevolution is merely long-term microevolution involves the functions of major organs like wings or complex traits like mimicry. The argument is that such major traits or systems could not have functioned in their present fashion during their postulated early stages. There are two obvious answers. First, many complex organs or adaptive systems almost certainly did have the same general function all during their evolution. Eyes are a good example, since even individual cells can become light-sensitive by slight changes in the plastids or pigment bodies present in their cytoplasm.

Mimicry is a similar example. Anyone familiar with the astonishing degree to which the traits of palatable or harmless butterflies or snakes resemble those of poisonous or venomous species (e.g., Wickler, 1968) could not be blamed for at least a momentary doubt that such extensive similarities could come about through series of small mutational changes saved by natural selection at every stage. Yet the process is simple to visualize. Consider a hypothetical pair of butterfly species, one palatable and one poisonous. Any slight tendency by a predator to confuse the two, and as a result to avoid both, would benefit the palatable species. Initial changes in the palatable species could be extremely small and subtle—such as whether it flew at the same times as the poisonous species, or how similarly it flapped its wings. The smallest color changes increasing resemblance could be powerfully selected. The inevitably increasing ability of the predator to distinguish the two would continue the evolution of even more elaborate mimicry. Changes of this general kind have in fact been observed directly in the generation-to-generation microevolution of cryptically colored moths in England, living in areas in which tree trunks on which they rest have, over decades, become alternately covered with dark soot and light speckled lichens (Ford, 1971).

Some major organs, like wings, must have served different functions in their early stages, even if fossils proving this hypothesis are unavailable. We can be sure that such kinds of change in function occur: the human arm and hand are surely derived from locomotory devices, and bird and bat wings from the forelimbs. For bird and bat wings it is only necessary to make the reasonable assumption that the forelimbs had become gliding aids before they were flying organs, as is the case today in so-called "flying" squirrels and lizards, neither of which actually flies, but rather only glides. In their earliest stages such structures may only have assisted the animal in safer landings during leaps between perches. Even before they were useful in gliding, some structures that later became wings may have functioned in courtship display, becoming elaborate enough in that context to initiate the evolution of the gliding function. This is the sequence sometimes postulated to explain insect wings, which are not derived from prominent organs like legs or arms with known functions prior to their role in flight (Alexander and Brown, 1963). Even without any fossil evidence there are often convincing reasons for accepting hypotheses of such functional shifts during evolution.

We should also not be surprised if some major organs, like insect wings, have evolved only once or twice, passing through their precursor stages rapidly and confounding us by leaving no extant intermediate stages and few helpful fossils. Thus, once some insects had evolved strong flying ability—especially aerial predators like the ancient dragonflies—the likelihood of other species passing successfully through the early clumsy stages, or remaining at some intermediate stage, may have been reduced to near zero. These kinds of happenings deny us the information we need to complete our understanding of certain evolutionary events, but they do not deny evolution.

The circle of reasonable argument and evidence on this crucial point is completed, again, by the ease of showing, by both selective breeding and hybridization, that major organs can be altered by individual gene mutations. Thus, evolutionary biologists invoke only known mechanisms to explain major organs, and there are excellent reasons for expecting these mechanisms to be wholly responsible for the appearance of traits whose entire, step-by-step evolution cannot be directly observed because of the brevity of human existence.

If microevolution does indeed lead to macroevolution, then to understand the latter we must analyze the former. More than this, if macroevolution is simply microevolution over long periods, then whatever forces operate in microevolutionary change have produced all of the traits of life by an accumulation of their effects. Should this be true, we ought to be able to use the cumulative effects of microevolution to develop a very large number of predictions about the nature of existing yet poorly known forms of life—predictions which can be tested by carefully selected comparisons.

Darwin's Comparative Method

Darwin's comparison of species on different islands exemplifies the kind of exploitation of naturally occurring situations or natural "experiments" referred to as the "comparative method" by modern biologists. It is a way of answering questions when human-designed experiments are not feasible—in this case, a way of reconstructing events that take so long they can never be observed directly by individual humans. Michael Ghiselin, in his book *The Triumph of the Darwinian Method* (1969), has noted that Darwin used the same method to develop a theory about the origin of coral atolls; that theory too has stood the test of time.

I think we must acknowledge that Darwin's comparative method is a scientific tool paralleling the microscope and telescope, and just as significant in broadening our perceptive horizons. Optical instruments and various indirect methods of chemical and physical analysis extend the range of our perceptive abilities in terms of the *space* occupied by *objects,* and, among its other values, the comparative method extends the range of our perceptive abilities in terms of the *time* occupied by *events.*

Comparative method is not unique to biology, but, leaving aside astronomy, it is less familiar in the study of the physical universe, and it is a part of the methodology of evolutionary biologists that makes evolution difficult to understand. Its essence lies in randomizing the effects of variables by exploiting diversity rather than in eliminating them by manipulation, as is commonly done in laboratory experiments. Its special virtues are its usefulness when manipulative tests are impossible, and the relevance of its results to the natural situation because it does not involve manipulation. The difficulty in making it precise and useful involves locating the diversity required to randomize the effects of confusing variables (see also Alexander, 1978a). Biologists are fortunate that life is so richly diverse and has evolved over such a long period; as a result it often provides the wealth of comparisons needed to answer evolutionary questions.

A good example of comparative method in everyday life is the collecting of data on the value of automobile seat belts. The definitive experiment would involve planned accidents with human subjects and cannot be conducted (for example, one might cause a large number of drivers to crash their cars at the same speed but allow only half to fasten their seat belts). Experiments with dummies and nonhuman primates have been useful, but the most convincing results have come from meticulous comparisons of unplanned accidents sufficient in number and diversity to allow randomizing of the unexpectedly large number of influential variables. There are differences in the kinds of drivers who fasten seat belts, differences in the kinds of cars purchased by people who do or do not fasten seat belts, changes in driving habits after starting to fasten seat belts, plus a great many others (Campbell et al., 1974).

Numerous examples of modern field studies in biology now exist, involving the comparative method and yielding the kinds of results important to arguments of the sort presented here: Howard (1979a, 1979b, in press), Hoogland (1977; in press, a, b, c), Sherman (1977, in press), Hoogland and Sherman (1976), Low (1978, 1979). Examples especially relevant to the topic of this book (and discussed later) include Alexander (1977b), Alexander et al. (1979), and Alexander and Noonan (1979).

A good illustration is the analysis by Alexander et al. (1979)

of the relationship between sex differences in size and the breeding system in mammals. In most mammals adult males are larger than adult females and these investigators discovered that in three different groups, ungulates, pinnipeds, and primates, this male-female difference increases as harem sizes increase. The probable selective force is that competition for females increases in intensity as some males achieve high success. This result occurs because sex ratios do not adjust to the breeding system so that, with the increasing success of a few males, an increasing number of males are reproductively disenfranchised. The increased competition among males causes larger males to be more successful than smaller ones, presumably because they can defeat smaller males in combat or competition for access to females. The results are persuasive for several reasons: (1) breeding systems and sexual dimorphism have evidently evolved separately in these three very distantly related groups, yet are closely parallel in all three cases, (2) the correlation exists whether mean or maximum harem sizes are used, and (3) the correlation even exists in two of the three cases if monogamous species are eliminated and only polygynous species with differing harem sizes are compared. Moreover, various studies have shown that larger males indeed are more successful breeders in polygynous mammal species (Le Boeuf, 1974; Kitchen, 1974).

These results clearly indicate that long histories of directional selection underlie male-female differences in mammals with different kinds of reasonably stable breeding systems, and they yield predictions for other situations not yet analyzed. There is no easy way, except by comparative study, to test most questions about the long-term history of life, or to generate predictions from evolutionary considerations.

I suspect that the comparative method, used both consciously and subconsciously, actually represents our main source of information about the universe. There can be no doubt that it will remain central in answering the kinds of questions that will be raised at an increasing rate by the evolutionary approach advocated in this book.

We may never know whether Darwin came first to the idea of long-term change and later sought a mechanism to account for it, or first realized the significance of natural selection and then sought evidence of its long-term operation. The crucial point now is that the predictiveness of evolutionary theory is not to be found by analyzing patterns of change across geological time, as evidenced by fossils. Instead, this predictiveness must be sought in the mechanism of change, the actual *process* of evolution that is directly observable in changes from generation to generation in the phenotypes—or structure, physiology, and behavior—of living organisms. Predictiveness with regard to evolution thus depends on our ability to develop testable theories about the long-term cumulative effects of natural selection. For example, even if genetic variations do not correlate with behavioral variations in different circumstances, the particular kinds and ranges of behavioral variation are usually understandable as cumulative effects of selection on developmental or phenotypic flexibility. This point is particularly important in a book about humans; later I will argue that an exactly parallel view must be taken of culture if we are to understand its evolution and develop a predictive theory about cultural change.

The whole reason for phenotypes' having evolved is that they provide flexibility in meeting environmental contingencies that are only predictable on short-term bases. Learned behavior is the ultimate of all such flexibilities. Not just humans and higher mammals but animals in general develop their behavior, or "learn," to do what is appropriate in their particular life circumstances. Even the remarkably distinctive castes of the social insects are in nearly all cases determined not by genetic differences, but by variations in experiences with food or chemicals while they are growing up. The ranges of variation, and the adaptive "peaks" along the axes of such variations (in the case of the social insects, the actual worker and soldier castes), are finite and predictable (e.g., Oster and Wilson, 1979). I believe we will eventually discover that exactly the same is true for the range and relative likelihoods of composites of learned behavior (or "learning phenotypes") in humans (Alexander, 1979a).

In meeting the challenge of Darwinism to understand ourselves, then, we must (1) discover precisely what are the forces of microevolutionary change, (2) determine the importance of each, and (3) learn how to predict their effects. Then we must ask the same questions about cultural change.

THE STRUCTURE OF EVOLUTIONARY THEORY

A theory is said to be a simple set of propositions that provides a large number of explanations. Einstein noted that "a theory is the more impressive the greater is the simplicity of its premises, the more different are the kinds of things it relates and the more extended is its range of applicability." Although he was not referring to evolutionary theory his statement could scarcely have applied more appropriately. For all that it purports to explain, evolutionary theory is based on a remarkably simple set of propositions. The process from which it stems derives from the interactions of five basic phenomena:

Inheritance: All living organisms (phenotypes) are products of the interaction of their genetic materials (genotypes) with their developmental (ontogenetic) environments; these genetic materials (genes, chromosomes) can be passed from generation to generation unchanged. Without inheritance there could not be cumulative change.

Mutation: The genetic materials do change occasionally, and these changes are in turn heritable. Without mutations there would be no continuing source of change (in forms lacking culture).

Selection: All genetic lines do not reproduce equally, and the causes of the variation may be consistent for long periods. Without selection there would be no direction to cumulative changes.

Isolation: Not all genetic lines are able, for various intrinsic and extrinsic reasons, to interbreed freely, and thus continually to re-amalgamate their differences. Thus, some populations cannot interbreed because they are spatially or temporally (extrinsically) separated; others are so genetically (intrinsically) differ-ent as to preclude hybridization. Without isolation there would be but a single species.

Drift: Genetic materials are sometimes lost through accidents, which are by definition random or nonrepetitive in their effects on populations. The main effect of genetic drift is to reduce the influence of selection, especially in very small populations; evo-lution could of course occur without drift.

These five phenomena have all been demonstrated repeat-edly, and they can be demonstrated at will, as can some of their interactions. No living things have been demonstrated to lack any of them, or are suspected to lack any of them. Hence, they may be described as the *factual basis* of evolution.

The *theory* of evolution, then, is the proposition that the effects and interactions of these five phenomena, in the succes-sions of environments in which organisms have lived, account for the traits and history of all forms of life. The challenge we face here is how to apply this simple proposition toward a better understanding of human sociality.

Of the five main components of the evolutionary process, natural selection, or the differential reproduction of genetic variants, is generally accepted as the principal guiding force. The reasons for this acceptance are not commonly discussed; it seems to me that there are at least three. First, altering the directions of selection apparently always alters the directions of change in organisms; this indicates that evolutionary change does not depend for its rate upon the appearance of mutations. Second, the causes of mutation and the causes of selection appear to be independent; and, third, only the causes of selec-tion remain consistently directional for long periods, and, hence, could explain long-term directional changes.

Mutations are caused, at least chiefly in the past, by atmo-spheric radiation or, perhaps, by internal chemical events still poorly understood (Suzuki and Griffiths, 1976). Selection, however, is caused by extrinsic phenomena that Darwin termed the "Hostile Forces of Nature": climate, weather, food short-ages, predators, parasites, and diseases. This list implies competition for resources, such as food and shelter from the other hostile forces. Accordingly, for all sexual species, we must include competition for mates as a selective factor. Darwin distinguished such competition from natural selection, calling it sexual selection, perhaps because he recognized that sexual competition could cause the spread of traits lowering resistance to the other hostile forces, therefore giving the appearance, at least, of making the species as a whole more vulnerable to extinction. Thus, large antlers may enable male deer or elk to mate with more females but simultaneously render them less able to escape speedy predators. This kind of selection may have seemed to Darwin to be inconsistent with the effects of selection on other kinds of traits, in which there is apparently a greater coincidence of usefulness to the individual and usefulness to the species.

The competition involved in natural and sexual selection is not just for the greatest quantity of resources but also for the highest quality. Those organisms will outreproduce that use the least energy and take the lowest risks in securing the highest quality and quantity of resources and converting them into their own genetic materials.

It is crucial to understanding the behavior of organisms, including ourselves, that in evolutionary terms success in reproduction is always *relative;* hence, the striving of organisms is in relation to one another and not toward some otherwise quantifiable goal or optimum. As we shall see later, this simple fact about biological conflicts of interest connects to the most profound enigmas of human existence. Thus, all across history the deepest thinkers have pondered, and never fully understood, why justice is necessarily incomplete, and why individuals in our highly social species are often prone to the feeling of being "alone in a crowd," or of never being fully understood. For such questions it is surely no triviality to comprehend that we have evolved to have *individually separate interests,* and to strive *in relation* to one another, because each of us is genetically distinct.

Because directions of mutation evidently are random with

respect to directions of selection, mutational changes as such are independent of adaptation, or the behavioral, physiological, and morphological fine tuning that organisms exhibit in response to their physical and biotic environments. The same is true of genetic drift, for by definition its causes are without cumulative directional effects on the genetic materials. This means, first, that as evolutionary adaptation proceeds, mutations must increasingly tend to become deleterious, so that their rates of occurrence have likely been selected severely downward. It also means that directional evolutionary change cannot result from either mutations or drift, but must be caused by directional selection. The only exception that I can imagine, and it is purely hypothetical, is the concept of selection suddenly becoming absent in the environment of a complex organism, with mutational changes then leading to steady reductions in complexity. This effect has sometimes been postulated when some particular selective pressure has evidently disappeared (e.g., reductions in size and complexity of human teeth with the advent of cooked food, or disappearance of eyes in cave animals). Such cases, however, are more appropriately explained as changes in directions of selection. In no way do they support an argument that selection itself somehow mysteriously disappeared from the organism's environment. When one direction or force of selection is removed from the environment of a species, the necessary effect is to cause other previously opposing forces to become more intense or powerful.

These are the reasons, then, for the common tendency to refer to the theory of evolution as the theory of natural selection; we derive them by applying logic to the set of facts known, from experiments and observations, about the phenomena which together make up the process of organic evolution. So we are led to the next conclusion: *to understand ourselves better from our evolutionary background we must focus our attention on one particular part of the evolutionary process—on the causes and effects of differential reproduction or natural selection.*

Darwin's Claim of Universality

Theories are supported by their ability to explain observations, and to explain them better than alternative theories. They are discarded when their predictions are at odds with verifiable observations. It is possible, therefore, to prove that a theory is incorrect, but, as suggested by earlier comments on Darwin's challenge regarding complex organs, it is difficult to prove in the same way that another theory is the correct one. To prove a theory correct it would be necessary to test all possible predictions. Since this is impractical, if not impossible, the correctness of a theory, especially a theory with great generality, can be supported but not proved.

This logical problem has sensitized scientists and philosophers to the importance of locating, for any theory, observations or predictions that if true would falsify it. The theory in question will be supported if such predictions repeatedly prove wrong, or if such observations are impossible because the falsifying conditions do not exist. Theories ought not to be regarded as useful unless means exist by which they can be falsified, whether through alternative theories or contrary predictions.

The claimed universality of Darwinian selection has caused a succession of writers to describe it as a nonfalsifiable theory —a useless theory which explains nothing because it explains everything. I have already suggested that such critics have failed to read Darwin carefully, and I have cited one of the challenges he formulated for his own theory. Here is another (Darwin, 1859, p. 201):

> If it could be proved that any part of the structure of any one species had been formed for the exclusive good of another species, it would annihilate my theory, for such could not have been produced through natural selection.

Thus, in his original publication of the theory of evolution by natural selection, Darwin himself identified different ways in which the theory could be falsified. In effect, he said that his theory, if correct, should explain everything *observable* but not everything *imaginable*. Moreover, he did not say that an exception to his view of adaptation would weaken or diminish his theory, rather that it would *annihilate* his theory. This is indeed a bold challenge.

Darwin employed still another method of testing his theory. He deliberately sought actual biological phenomena that seemed most difficult to explain by natural selection. Thus he said of the sterile castes of social insects that they presented "the one special difficulty, which at first appeared to me insuperable, and actually fatal to my whole theory." Considering that Darwin was unaware of the existence of genes, or their distribution among related individuals, his solution to the problem of sterile castes was uncannily accurate. He noted that traits may be carried by members of the family that never express them, and if the expression of those traits contributes to the reproduction of family members who are not expressing the traits but nevertheless carrying them, then the traits themselves can be advanced by natural selection. As he put it (Darwin, 1859, p. 238), "a breed of cattle, always yielding oxen [castrates] with extraordinarily long horns, could be slowly formed by carefully watching which individual bulls and cows, when matched, produced oxen with the longest horns; and yet no one ox could ever have propagated its kind." Similarly, he noted that tasty vegetables could be produced by saving seeds from relatives of the vegetables tasted or eaten, therefore unable themselves to produce seeds, and cattle with "the flesh and fat . . . well marbled together" could be bred although "the animal has been slaughtered" if "the breeder goes . . . to the same family." Thus, he solved the worst problem he could locate for his theory of natural selection by realizing that the *trait of sterility* could be favored if it contributed sufficiently to the reproduction of close relatives or family members. In other words, sterile helpers could perpetuate the tendency for individuals in certain kinds of helping situations to become sterile if they thereby contributed enough to the reproduction of the family members they were helping (see also Fisher, 1930; Haldane, 1932; Williams and Williams 1957; Hamilton, 1964). Since we know now that selection proceeds when variations in traits correlate with variations in genes, Darwin's analysis was remarkably prescient, actually anticipating the theoretical refinements of evolution responsible for this book.

By demanding that the traits of organisms take their forms largely because of their selective backgrounds, Darwinian theory rejects the possibility of certain kinds of altruism (or beneficence) being adaptive or reproductively advantageous. Although Darwin spoke only of "structure" in this connection, we are obviously forced to expand the challenge to include all traits, whether morphological, physiological, or behavioral. Although he spoke only of altruism *between* species, we cannot overlook the additional fact that all forms of genetic or reproductive altruism (table 1) *within* species are also contrary to evolutionary theory, and should exist only as a result of accidents, or suddèn environmental changes which leave an organism temporarily maladapted or incompletely adapted. Whenever present, such altruism should tend to diminish (however slowly in relation to other processes like cultural change) until it disappears completely. Evolutionary theory may seem mundane when one is discussing its general structure, but its consequences, especially in regard to the everyday concept of altruism, are nothing short of astonishing (Note that I am here using altruism to refer to the consequences of acts and have not yet introduced questions of motivation or intent).

The problem of falsifying evolutionary propositions still looms large in the minds of some skeptics, such as philosophers of science who are well versed in the methodologies of the physical sciences yet who perhaps know little about the biology of organisms. Accordingly, I will direct special attention to the nature and significance of both general and particular tests of evolutionary questions. I caution the reader, however, that there is no single, overall test that is both simply applicable and able to falsify the entire theory of evolution in all its aspects; nor should one necessarily be expected or required. For a process that has led to the diversity and complexity of the life existing throughout geological time, the only reasonable expectation is that we must repeatedly match our predictions with evidence and judge the theoretical significance of those predictions on the basis of their likelihood, individually and collectively, of being met by accident or chance. It is ironic that the biologists who accept evolution as a fact and use it successfully to study life do so because they have personally accumulated knowledge of many such tests, while the skeptical nonbiologists, who by definition are not aware of any such cumulative evidence, continue to seek the simple, once-and-for-all falsifying operation, and to deny or ignore evolution because they cannot find it. Such skeptics often demand that adaptation be defined in some abstract or general sense which could not derive from a process of differential reproduction (e.g., Stent, 1978). As Williams (1966) pointed out, however, natural selection does not lead to adaptedness in any absolute or universal sense, apart from the environmental context; instead it is merely a matter of better versus worse—or *relative* reproductive success —in the immediate environment.

One more thing needs to be said about the supposed circularity or tautology of the phrase "survival of the fittest." If we never could predict differential survival or reproduction, but could only analyze it in retrospect, this criticism would be justified. Of course, this is not so. We can make countless accurate predictions from variations in the attributes of organisms, such as, in an environment including sharp-eyed hawks and a white sand substrate, white mice will outreproduce black mice. Thus, *the concept of natural selection does not require circularity.* Predictions that prove this point repeatedly are scattered throughout this book and involve many kinds of attributes.

SURVIVAL OF THE FITTEST WHAT?

Now we have reached a point where Darwin's statements about evolution can no longer help us. We are faced with a crucial question that Darwin did not answer, and which I believe has been more responsible than the seeming nonfalsifiability of evolutionary theory for the disrepute into which evolution has at times fallen, and for the failure of evolutionary biology to answer questions satisfactorily during the first two-thirds of the twentieth century.

The question Darwin failed to answer was actually a simple one. Survival of the fittest *what?* The answer, which is not so simple, requires looking at the attributes and interactions of different units in the hierarchical organization of life. For Darwin this axis included traits, individuals, families, social groups, and species. For modern biologists there are more units, because genes and their various interactions and linkages, including chromosomes, have been added at the lower end. Among these units there may be conflicts of reproductive interest: what is best for the individual's reproduction is not always best for that of the gene or gene group, and what is best for the species does not always maximize the individual's reproduction. To understand how natural selection works, it is essential to discover the levels at which selection is most potent. In other words, when there are conflicts among these different levels, do traits exist because they help genes, individuals, populations, or the species as a whole? Darwin could not answer this question, and there are some indications that he recognized the problem and deliberately sidestepped it. Thus, he remarked (Darwin, 1871) that the perplexing problem of how to explain sex ratios (proportions of males and females), which gives the impression of being a population-level phenomenon, might better be left for the future. On the other hand, Darwin (1859) recognized that the ability of two populations to produce fertile hybrids if their members crossbreed cannot be eliminated by selection, since *individuals* that produce *fertile* hybrids will outreproduce those producing *sterile* hybrids. There can be selection against the wasting of calories and the taking of risks in cross-mating between individuals belonging to different species, but not toward lowered fertility if such matings occur, even if the effect on either or both *species* (as a whole) is detrimental. As already noted, Darwin's treatment of sexual selection also implied puzzlement about the units of selection.

One has to conclude from extensive reading of Darwin that he was uncertain about the identity of the units of selection. For almost one hundred years the question of the potency of selection at different levels was not even clearly posed, and its resolution has proved so complex that many aspects remain to be analyzed (Leigh, 1977; Dawkins, 1977; Alexander and Borgia, 1978). Nevertheless, a general consensus among evolutionary biologists is clear: adaptiveness is not appropriately assumed at any higher level of organization than necessary to explain the trait in question (Williams, 1966). This means that genes and their simplest groupings—about which Darwin had no knowledge—are the usual units of selection. Here I will use three topics to illustrate how this consensus has developed since about 1957.

POPULATION REGULATION, SEX RATIOS, AND SENESCENCE

Population Regulation

The British ecologist David Lack began to answer the question of the relative significance of adaptiveness at individual and population levels when he took up the problem of population regulation (Lack, 1939, 1954, 1966), long recognized as central in the broad field of ecology. Lack showed experimentally that birds with small clutches could very well be maximizing their numbers of fledged offspring by giving more parental care to each nestling. There is no reason to suppose, as many ecologists have (e.g. Wynne Edwards, 1962), that such birds are keeping their clutch sizes low in an altruistic effort to regulate their populations at optimal levels. Lack's argument is theoretically satisfying, since it would seem that any birds foregoing the maximization of reproductive success would be outreproduced and replaced by more "selfish" birds.

Lack's findings implied that among organisms in general, those which have evolved to produce smaller and smaller brood sizes, or as with humans to produce single offspring less and less frequently, might thereby be increasing their actual reproduction because of the value of the additional parental care they could give to the few offspring produced. Interestingly enough, Darwin (1871) had anticipated this finding. He remarked that the Fulmar Petrel, though it lays but one egg, was then thought to be the most abundant bird species in the world. More generally, he commented as follows on population regulation:

> The only check to a continued augmentation of fertility in each organism seems to be either the expenditure of power and the greater risks run by parents that produce a more numerous progeny, or the contingency of very numerous eggs and young being produced of smaller size or less vigorous, or subsequently not so well nurtured. [Vol. 1, p. 319]

Darwin was saying that in the race of natural selection, individuals are kept from evolving to produce more and more offspring by (1) forces preventing increases in size (hence, preventing the possibility of having more benefits to give), and (2) the necessity of either giving fewer parental benefits to each offspring, or taking greater risks in the process of giving those benefits. Darwin's statement embodies the principle that reproductive effort (the expenditure of calories and taking of risks in the act of reproduction) is finite, that it evolves continually to be fully and most effectively expended, and that it is somehow more effective at the individual level than at any higher level. The clear implication is that, in evolutionary terms, the lifetime of an organism is no more and no less than a strategy of reproduction by a group of genes—the *genotype* or *genome* of the individual.

It is also clear that we must expect all functions of the organism to be reproductive, and maximally so. Darwin's argument amounts to a hypothesis that every individual's lifetime is a series of unwitting cost-benefit decisions, leading, in the organism evolutionarily adapted to its environment, to a maximization of *reproduction* of its genetic materials; this reproduction in turn apparently maximizes the likelihood of *survival* of the various units of genetic materials. (I must say here that even if this has been true of humans in the past, it is not necessarily so for the future because of our evolutionarily novel ability to reflect consciously about personal motivations, and to ask questions about our background and seek answers to them; I shall have more to say on this later.)

The ramifications of this rather simple idea about genetic reproduction are astonishing. It seems ludicrous to suggest that all activities of humans derive from the reproductive strivings of individuals, or more properly their genes (except activities that because of environmental changes are temporarily maladaptive—in the biological sense of not being maximally reproductive). Unless there are flaws in the argument presented so far, however, we are compelled to examine this hypothesis.

Darwin's statement about fertility, quoted above, may be regarded as a prediction. To my knowledge the only animal group in which an extensive effort has been made to test it is amphibians (Low, 1976). Figure 1 reproduces the results, which indicate unmistakably that where egg size is large only small clutches are found, and that with respect to small eggs the larger clutches are produced by larger females. Darwin's statement implies that selection is universal and inevitable; thus, it actually points to another possible way to disprove evolutionary theory, and Low's comparison is an example of such an effort, the results of which support rather than falsify Darwin's argument (therefore, again giving it credence against imagined alternatives like divine creation).

The usefulness of Low's broad (but necessarily superficial) comparisons could be increased by the identification of pairs and groups of closely related species, which should conform more closely than average to Darwin's prediction because confusing variables are likely to be minimal among similar forms. Likewise, more precise clutch sizes may someday be known. Paralleling Low's findings, Cowan (1978) has shown that within certain species of parental wasps the adult's size is determined by the amount of food furnished by the mother, and larger mothers both provide more food and have more young. That only such rudimentary tests of Darwin's argument exist

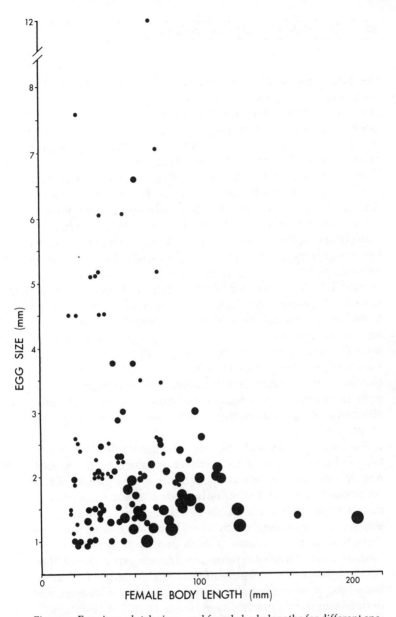

Figure 1. Egg sizes, clutch sizes, and female body lengths for different species of Amphibia, showing the general correlations that Darwin suggested:
bigger females have more to invest in eggs; and for females of a given size,
clutch size and egg size correlate inversely. The five sizes of circles represent
clutches ranging from less than 500 to more than 10,000. Each dot represents
a different species. (Modified from Low, 1976.)

attests, I think, to our failure to recognize the profundity of many of his conclusions.

Before the work of Lack and Williams, ecologists had several reasons for thinking that population regulation occurred as a group-selected trait. First, they had observed that under natural conditions populations do not ordinarily fluctuate violently in size, and they realized that such stable populations were less likely to overuse their resources so as to crash and run the danger of extinction. Second, they observed that population size commonly levels off before food and other evident resources are utilized entirely, and that this sometimes occurs because (1) brood or clutch sizes are lowered at high population densities or (2) some proportion of adults fails to breed at all at such densities. Third, ecologists were misled because a major source of *extrinsic* regulation—predation—had often been modified by man; for example, it seems likely that major predators on every widely studied rodent (the group of animals probably most prominent in studies of population cycles) have been reduced by human presence. Moreover, predation is often difficult to observe even when it is known to be important. Finally, so long as no one gave reasons to doubt theories of intrinsic regulation of a group selection nature, such theories were easy to accept.

It is curious that an obvious test of a group-altruistic theory of population regulation seems never to have occurred to anyone. It concerns breeding patterns of organisms that reproduce only once in their lifetimes. Individuals of iteroparous (repeat-breeding) species—such as most vertebrates, including ourselves—may gain reproductively by reducing or eliminating reproduction during times of high density or scarcity of resources; by avoiding risks or needless caloric expenditures, they may reproduce more in better times. Individuals of semelparous organisms (onetime breeders), such as salmon, soybeans, and most insects, could not realize such a benefit. The test, then, is to locate *a semelparous organism that withholds reproduction when its population is very dense* and that also has no possibility of helping nondescendant relatives reproduce. The existence of any such species would be a case for intrinsic population regulation, genetic altruism, and group selection. No such case has been reported. Similarly, if one could locate *a semelparous species without parental care in which individuals reduced their brood sizes at high densities* (beyond what might be caused by immediate shortages of resources convertible into offspring) this would support the idea that population regulation is intrinsic and depends upon a group-selected altruism. In fact, unless selection among individuals rather than populations is responsible for the existence and nature of the traits of organisms, the above, easy-to-identify traits should be prevalent among the tens of thousands of semelparous organisms. Accordingly we have located still another test that would falsify the theory of evolution by natural selection, at least insofar as it operates principally through the differential reproduction of individuals. Unlike some of the other tests of evolutionary theory, this one seems easy to perform.

Sex Ratios

Sir Ronald A. Fisher (1930) responded to the "levels of selection" question by developing essentially all of the arguments in his book, *The Genetical Theory of Natural Selection,* in terms of benefits to individuals. The only exception was his discussion of sexuality, and in the context in which he developed it that may have been appropriate (Alexander and Borgia, 1978; Maynard Smith, 1978; Leigh, 1978).

One of the most effective of Fisher's arguments for the potency of selection among individuals was that, when sexual competition is population-wide, population sex ratios derive from individual sets of parents evolving to produce sex ratios in their broods such that parental investment in the two sexes is equal in the breeding population as a whole. In effect, a parent should produce a sex ratio in its brood which, in the local environment of sexual competition, does not lead to the reproductive devaluation of any of the offspring on account of its sex. Generally speaking, in a population in which all individuals of one sex are potentially available for mating to any given individual of the other sex, if the sex ratio in general is one male for one female, and the two sexes of offspring are equally expensive to produce, no parent can do better than to produce a one-to-one sex ratio in its brood (for exceptions see Trivers and Willard, 1973; Alexander, 1974).

Fisher's argument stipulated that sex ratios at conception would not adjust to the breeding system in the particular way that a group selection hypothesis would require. In other words, if some males have large harems it does not follow that the number of males produced in the group or population will be correspondingly reduced, even if this would be to the population's advantage (for a detailed explanation, see Alexander and Howard, in prep.). Numerous observations confirm that such compensation in fact does not occur. A consequence, already mentioned, is that the intensity of sexual competition varies between the sexes as breeding systems deviate from monogamy. Thus, if sex ratios do not adjust to the breeding system, then if some males have harems, others have no mates at all. Many structural, behavioral, and other differences between the sexes are explained by the variations in competitive circumstances deriving from this simple observation (e.g., see below and Alexander et al., 1979).

Hamilton (1967) contributed a major supplement to Fisher's theory of sex ratio selection by noting that when local competition for mates is more intense in one sex than in the other, as when females are inseminated by their brothers and one male can inseminate several females, then parents should invest less in the sex that is devalued reproductively by the competition for mates (except that in a monogamous species in which males and females invest equally in the offspring even brother-sister matings would not produce such biases). Hamilton cited many cases in which broods inbreed and are heavily female-biased, and since then additional evidence has accumulated. Among related species of bees and wasps, for example, females that nest alone are more likely to bias investments in the sexes than those that nest in colonies; this is as predicted, since broods reared alone are more likely to engage in brother-sister mating.

Individual females within species may also be able to adjust sex ratios to fit either situation (Cowan, 1978; Alexander and Sherman, 1977).

Although we know little about how females can adjust sex ratios it is apparent that they do; thus, in humans and other mammals it seems certain that far more males than females are conceived, and sex ratios approach one-to-one only some time following birth (McMillen, 1979). Several possible mechanisms exist, such as (1) differential mortality of the two kinds of sperm in the reproductive tract, (2) differential penetrability of eggs to sperm bearing X and Y chromosomes, and (3) differential elimination of embryos of the two sexes. In the Hymenoptera (such as bees, wasps, and ants) and some other insects, males result from unfertilized eggs, females from fertilized eggs; the female stores the sperm after mating in a sperm sac called the spermatheca and she is able to determine the sex of each offspring individually by whether or not she fertilizes the egg as it is laid.

These results with sex ratios are alone sufficient to suggest that the traits of organisms result from natural selection effective at no higher than the individual level. In other words, data on one set of attributes of living organisms may demonstrate both the universality of evolution and the mode of its operation. They do this because of the generality and testability of the predictions about the cumulative effects of selection on the single attribute of sex ratios. Fisher's sex ratio predictions were a major early example of a class of highly predictive phenomena that Maynard-Smith and Price (1973) called "evolutionarily stable strategies," meaning "strategies" for which, once they are widely adopted, there is no more reproductive alternative. Evolutionarily stable strategies are, in turn, one kind of predictable, long-term, cumulative effect of natural selection.

Senescence

George C. Williams (1957) made an early contribution to the modern refinement of Darwinian selection by developing the only extant viable theory to explain senescence and the finiteness of individual existence. He argued that if genetic evolution has led individuals to maximize their reproduction, then genes having beneficial effects early in life will be saved by selection even if they have deleterious effects later in life. Such genes will be saved because their action early in life causes them to help more individuals (because fewer individuals will have died) and more of the reproduction of each individual (because at a younger age more will be left of each individual's reproduction). The consequence of selection for such genes is an accumulation of late-acting, deleterious effects and an ever-increasing likelihood of death. Although selection will work against such effects, so long as they inevitably accompany earlier-acting beneficial effects of the same genes they can never be entirely eliminated; instead the deleterious effects will tend to be crammed together near the end of life when the reproductive value of the individual is minimal and selection is, therefore, least effective in counteracting them. Williams' theory had been suggested in part, and less precisely, by Medawar (1955, 1957), and it was later substantiated and expanded by Hamilton (1966). It is centrally important as the first creditable theory by which the general programming of life cycles and the finiteness and variety of individual existences could be understood as a result of selection for individuals who maximize their own reproduction (see also Lamb, 1977; Alexander and Howard, in prep.). Without this theory it was commonly supposed either that natural selection led to increasing likelihood of long-term survival of individuals or that individuals senesced and died as a result of an altruistic or group-selected tendency to make way for younger, healthier members of the species. The first of these ideas is inconsistent with the fact that in the vast majority of species individuals have very short life cycles. The second has no possibility of explaining either the remarkable variations in life cycles, from a few minutes in some microorganisms to several thousand years in redwood trees and bristlecone pines, or the distribution of life cycle differences within groups of closely related species.

Williams' theory of senescence and the finiteness and general

brevity of individual existence, together with observable variations in longevity among species, and variations in patterns of
activities and susceptibility to various sources of mortality during the lifetimes of individuals within species, represent a particularly powerful source of support for the effectiveness of
selection at low levels in the hierarchy of organization of life.
Human mortality curves, for example, are astonishingly similar
for widely different times and places, and the mortality curves
of the two sexes tend to maintain a characteristic relationship
to one another (figure 2). This relationship is approximately
what one expects for a mildly polygynous species (i.e., one in
which the variance in male reproduction is higher than that in
females—Trivers, 1972), an empirical fact for the human species. The hypothesis that selection favors the persistence of the
genes of different individuals rather than the entire gene pools
of populations leads to a number of predictions about differences between the sexes in a polygynous species; all of these
predictions are true of the human species (Alexander et al.,
1979, Alexander, 1978d; Alexander and Noonan, 1979):

Males are expected to be larger than females, because males
compete harder for females than vice versa, and because they
fight over females (for exceptions and the reasons, see Alexander et al., 1979). Males also are expected to possess special
means other than size and strength, such as weapons and ornaments, for competing directly with sexual rivals.

Males are expected to be conceived and born in higher numbers than females, because, as investment by parents, they are
higher in both risk and potential genetic return than females;
the expected strategy is to start more males and save only those
most likely to be successful.

Males are expected to suffer higher mortality rates than
females throughout life, and to senesce more rapidly. This is so
because their more intense sexual competition directly increases
mortality of males, and because higher mortality lowers the
value of beneficial gene effects late in life, causing an evolutionary acceleration of the severity of senescence (Williams, 1957;
Hamilton, 1966).

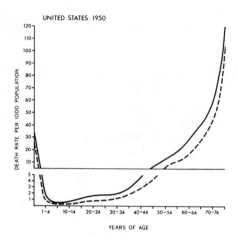

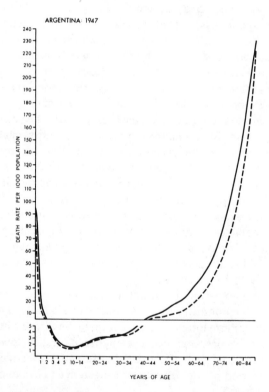

Figure 2. Lifetime mortality curves for human males (solid lines) and females (dashed lines) from different societies. The general shapes of the curves remain the same across the world, and the differences between the sexes also remain about the same. The implication is that human lifetimes follow predictable

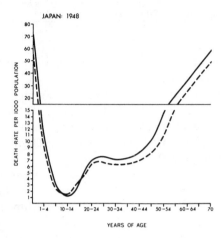

JAPAN: 1948

DEATH RATE PER 1000 POPULATION

YEARS OF AGE

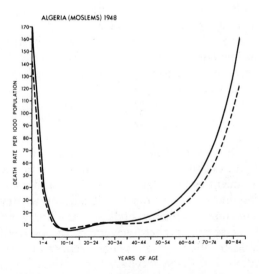

ALGERIA (MOSLEMS) 1948

DEATH RATE PER 1000 POPULATION

YEARS OF AGE

programs, evolved as a result of natural selection. (Data from *United Nations Demographic Yearbook*, 1954 and 1957; and from U.S. Department of Health, Education, and Welfare, National Office of Vital Statistics, Special Report 37, 1953.)

Males actually reared by their parents (as opposed to being aborted or otherwise terminated prematurely) are expected to receive more parental care than females (they are in fact carried longer in utero and are larger at birth); this prediction follows from the likelihood of greater reproductive success of most successful males as compared to most successful females (Bateman, 1948; Trivers, 1972).

Males are expected to have longer juvenile periods than females, because it is more difficult for them to enter the breeding population, and additional juvenile life gives time for extra growth and experience.

Healthy and socially high-ranking parents are expected to produce broods that are more male-biased than those of less healthy and lower-ranking parents (Trivers and Willard, 1973; Alexander, 1974; Dickmann, 1979), because they are more likely to be able to make highly successful male offspring.

Sex and gender "roles" (behavior) are expected to be more crucial (i.e., sex-typing more rigid) for male offspring than for female offspring (Luria, 1979; Alexander, 1979e; Alexander, 1979d), again because of the greater difficulty likely to be experienced by males in entering the breeding population.

Obviously, the possibility of all of these predictions being met by chance is vanishingly small. Equally obviously, these predictions involve facets of human existence for which explanations are at least interesting and in some cases possibly of great practical significance.

GROUP SELECTION

In 1966 Williams published a book criticizing what he called "some current evolutionary thought," in which he set out the general argument against "group selection" and chastised biologists for invoking selection uncritically at whatever level seemed convenient. Williams' book was the first truly general argument that selection is hardly ever effective on anything but the heritable genetic units or "genetic replicators" (Dawkins, 1977) contained in the genotypes of individuals.

Despite lingering arguments to the contrary (e.g., E. O. Wilson, 1975, p. 30), we can now say with conviction that Williams' argument against group selection was right. Lewontin (1970) began a systematic explanation of why this is so. He noted that, generally speaking, selection is more effective, or evolution more rapid, for units whose variants are more heritable, have more rapid generation or cycling time, and have greater variability among them. These attributes tend to characterize genes, chromosomes, and individuals rather than populations and species (figure 3). They also interact complexly, since greater heritability and rapid generation time both tend to reduce variability. Individuals in sexual organisms are the least heritable of the three units. Nevertheless, because of the complexity of individuals, the temporary integrity of the genotype during the individual's lifetime, and the potential of ontogeny (development, experience) to magnify in the phenotype the effects of genotypic variations, individuals are also the most variable of the three units. Selection, then, may be described as a consequence of differential reproduction of the genes of individuals, effected by different phenotypic performances of individuals, and it is evolutionarily (genetically) significant whenever differences in performance owing to genetic differences correlate with differences in reproductive success.

Usually, in considering whether a trait can spread, one may profitably ask what would be the fate of a gene imparting that trait to its bearers within groups containing other individuals that lack it. The potential error in this approach is failure to take sufficient account of the complexity and multiplicity of genic interactions, including their conflicts of interest. For example, Dawkins (1977), in a provocative paper arguing that we must see the world in terms of the interests of genetic replicators, does not stress the enormous problem of weighing conflicts of interest among the countless numbers and combinations of such replicators, yet this is the one problem in biology that becomes immensely more difficult if selection is only effective at low levels.

HIERARCHY OF UNITS OF LIFE			RATES OF PRODUCTION AND MORTALITY	GENETIC VARIABILITY AMONG UNITS	HERITABILITY OF FITNESS DIFFERENCES
ECOSYSTEMS AND COMMUNITIES			O	O↔●	O
SPECIES			O	O↔●	O↔●
POPULATIONS	long-lasting habitat	close together	O	O	O
		far apart	O	●	●
	short-lived habitat	close together	●	O	O
		far apart	●	O↔●	O↔●
SEXUAL GENOTYPES	inbreeding; much phenotypic plasticity		●	O	O
	inbreeding; little phenotypic plasticity		●	O	●
	outbreeding; much phenotypic plasticity		●	●	O
	outbreeding; little phenotypic plasticity		●	●	●
ASEXUAL GENOTYPES	between clones		●	O↔●	●
	within clones		●	O	●
CHROMOSOMES	sexual forms		●	●	●
	asexual forms		●	O↔●	●
SUPERGENES	sexual forms		●	●	●
	asexual forms		●	O↔●	●
GENES			●	O↔●	●

O   Low
O↔● Variable
●   High

Figure 3. Units in the hierarchy of life and their likelihood of being the units of selection. Black circles denote attributes enhancing the intensity of selection at that level. Sexual genotypes probably have the greatest amount of genetic variability, enhancing the potency of selection among them. But they are also the shortest-lived of all the units in the hierarchy. As Williams (1966) noted, genes are the most persistent of all living units, hence on all counts the most likely units of selection. One may say that genes have evolved to *survive by reproducing,* and they have evolved to reproduce by creating and guiding the conduct and fate of all the units above them in the hierarchy, in courses of action predictable only from knowledge of the environments in which they occur.

If, on the other hand, one attempts to answer questions about selection solely by considering the reproductive success of individuals, the most likely error is a failure to account for all contributions by individuals to copies of their own genes which may exist in fractional proportions in other individuals. Dawkins (1976) and Blick (1977) exposed such an error on my part (Alexander, 1974) in analyzing parent-offspring conflict. Blick showed how a gene in an offspring may spread itself even if its effects are contrary to the genetic interests of the parent and the rest of its brood. In Blick's example a mutant allele (or alternative form of a gene) leads its bearer to take from siblings resources that would have increased their likelihood of survival to reproductive age. The individual carrying the allele for such "robbing" behavior converts the stolen resources into an increase in its own likelihood of survival. One can consider three classes of outcomes: if the selfish juvenile converts the stolen resources into a greater number of grandchildren for its parent than would have been produced by a brood containing no selfish individuals (i.e., the conversion is more than 100 percent effective), the parent, its selfish baby, and the genes for robbing all gain. If the conversion is less than 100 percent but more than 50 percent efficient (i.e., the resources produce more than half as many, but not as many grandchildren for the parent as they would have if they had remained in a nonselfish offspring) then all of the genes in the selfish offspring's genome, including the allele for robbing, gain but the parent suffers a reproductive deficit (see Trivers, 1974). If the conversion of stolen resources is less than 50 percent efficient (in Blick's example, it is only 25 percent efficient), then the allele for the selfish behavior gains but other genes in its own genome do not, because each of them also has a 50 percent chance of occurring in the genomes of the robbed siblings (Trivers, 1974). This means that any gene in the selfish individual's genome which partly or entirely suppressed the effect of the allele leading to robbing of siblings would thereby aid its own spread. Alexander and Borgia (1978) have termed genes that act against the interests of other genes in their own genome "outlaw" genes and argued that their chances of suppression by mutant modifiers are very high because of the large number of genes in most genomes (tens or hundreds of thousands) and therefore the high chance that mutant suppressors would appear.

TABLE 1
Categories of Social Behavior

Genotypically	Phenotypically	Examples
selfish	selfish	engagements in reciprocity that on average lead to personal gain
selfish	altruistic	ordinary parenthood and nepotism
*altruistic	selfish	forgoing of both parenthood and nepotism
*altruistic	altruistic	adoption of an individual without known relatives by another individual without known relatives (at any time)

Note: Those categories asterisked will not evolve, and when they appear will tend to be diminished by selection. All nonsocial behavior is in the first category; most social behavior is in the second. The reason for requiring that neither the adopter nor the adoptee in the fourth category has known relatives is that the adopter could profit reproductively by help given to it or its relatives by either the adoptee or the adoptee's grateful relatives.

It is easy to imagine traits that would be genetically altruistic (table 1) such as lowered rates of reproduction that help the population survive but reduce the reproductive success of their individual bearers. For such traits to spread by helping the population, differential reproduction of populations within which such altruism was more prevalent would have to exceed that of populations within which it was less prevalent. Social groups and populations geographically close enough to replace one another in the event of unilateral extinction, unlike in-dividuals, characteristically lack the integrity to develop large differences that would result in differential reproduction. To support the concept of group selection it is not enough that groups or populations simply become extinct differentially, even if such groups also possess genetic differences, as they almost always will. For differential extinction of groups actually to be selective in nature, or capable of causing directional changes in the genetic materials (as opposed to extinctions owing to series of random changes or "drift"), the differences between the groups must be responsible for the differential extinction. This stringent requirement also leaves aside the difficult question of how altruistic traits spread when they first appear so that they can become more frequent in some popula-tions than in others and produce the hypothetical kinds of situations in which group selection seems to have some likeli-hood of operating. One can imagine that such traits could spread initially by genetic drift, or that changes in ecological circumstances could cause previously nonaltruistic traits sud-denly to become altruistic after becoming widespread; but such requirements are so stringent as to make group selection even less likely to be important. For these various reasons, any sig-nificant amount of evolution by group selection seems to be an exceedingly remote possibility (Alexander and Borgia, 1978).

The accumulation of supporting observations, and the failure of challenges like those just described, have led evolutionists to a general consensus that the evolved function or reason for existence of any trait cannot easily be ascribed at levels higher than that of the reproductive interests of the individual. The popular press and common belief notwithstanding, then, blue-jays do not scream to warn other species of the approach of predators; lemmings do not drown themselves to save their species from overpopulation; and parents do not adjust their brood size or the ratio of males and females in their broods to what is optimal for the social group, population, or species. Instead such things are evidently adjusted by natural selection toward maximization of the likelihood of survival of some genes, groups of genes, or chromosomes over their alternatives. This maximization results from the reproductive striving in-duced in individuals by the effects of these genetic units on the phenotype due to their actions in the particular environments in which the individuals develop and behave. Evidently blue-jays are either helping themselves when they scream (for exam-ple, by informing the predator of their awareness of it, thus discouraging it from continuing to hunt in the area), or they are warning mates or relatives of the danger. (For a review of possi-ble functions of "alarm" calls, see Sherman, 1977). Lemmings live in temporary habitats and must often cross fjords and streams to locate suitable new habitats; so it is not surprising that they enter water readily, or that they sometimes take very great risks and die in large numbers, in moving to new places to breed; one predicts that, in the situations in which such behavior occurs, to stay in the old resource-depleted or disease-ridden habitat would be even less reproductive. Parents appear to be producing the largest broods they can rear successfully, with male-female ratios that are most likely to maximize their numbers of successful grandchildren. The list can be continued. Until data are available to test statements like those above they remain as predictions, not conclusions; but their mere existence in this context has already revolutionized research in evolution-ary biology and led to innumerable new and supporting discov-eries.

The above arguments about individual reproductive "selfish-ness" (i.e., selfishness in terms of consequences if not intent) have proved very difficult for humans to accept. Biologists and nonbiologists alike have for the most part interpreted the activities of nonhuman organisms in terms of their value for the survival of the species, or at least for the good of the social group. This attitude probably derives from our tendency to anthropomorphize animals—to see them in our own image. It is a curious and startling fact that despite all of our talk about how, as individuals, we tend to look out for ourselves, scarcely anyone actually thinks of himself as behaving chiefly or solely so as to maximize his own welfare, much less his own reproduction. Instead, we cherish the notion that we are perfectly willing to behave unselfishly when the occasion calls for it; or at least we cherish the notion that we project such an image among our friends and kin. Exactly what this means is another question; in any case, it accords with our inclination to accept the widespread existence of altruistic behavior in nonhuman animals, or to defend its likelihood, whenever it seems a convenient explanation for a perplexing behavior.

THE CONCEPT OF INCLUSIVE FITNESS

Destruction of the notion that selection commonly results in attributes having evolved because they contribute to the welfare of the population as a whole raised questions about all behaviors that seem to be altruistic. Almost at the same time that group selection was being questioned by Williams, the alternative explanation for apparent altruism was also being generated. In two classic papers published in 1964, the British biologist William D. Hamilton developed a concept that he called "inclusive fitness" to distinguish it from what he thought of as "Darwinian" or "classical" fitness.

Inclusive fitness is a simple idea. It can be explained by reference to a diagram of genetic relatives and the nomenclature commonly applied to them within our own society (figure 4). As humans, we are sexual and social organisms. As sexual organisms we are able to reproduce only by helping or being helped by others; we do not fission and produce potentially immortal daughter cells. As social organisms we tend to lead our lives imbedded in networks of near and distant kin. The

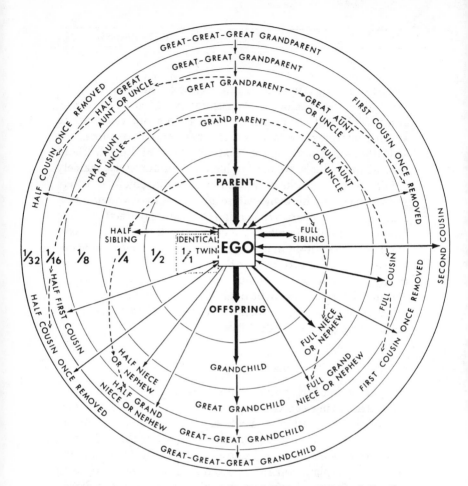

Figure 4. Genetic relatives potentially available to an individual, Ego, for reproductively self-serving nepotism. Arrows indicate likely net flows of benefits. Half the genes of parent and offspring are identical by immediate descent. Other relationships are averages. Dotted lines indicate closest relatives other than Ego, thus the most likely alternative sources of nepotistic benefits. Widths of lines indicate likely relative flows of benefits to or from Ego, based on the combination of genetic relatedness and ability of recipients to use the benefits in reproduction. Extreme lateral relatives are less likely to be encountered or identified because of social or geographic distance, extreme vertical relatives because they are less likely to be alive at the same time. Double-headed arrows indicate relatives whose status in regard to dispensing or use of benefits is doubtful owing to the uncertainty of their age relationships. (Thus, one's second cousin may be much younger, much older, or about the same age; one's sibling, on the other hand, is much more likely to be of comparable age.) Relatives on the right side of the diagram are those resulting from monogamous marriages; polygyny results in relatives indicated on the left. (From Alexander, 1977b.)

concept of inclusive fitness simply tells us that *not merely our offspring but any genetic relative socially available to us is a potential avenue of genetic reproduction.*

Fisher (1930) and Haldane (1932) evidently first discussed and quantified in rudimentary ways the idea that Hamilton was to develop in detail a third of a century later. Fisher wished to explain how distastefulness can evolve in caterpillars. If he had accepted that traits may evolve because they help the group, even if they diminish the reproduction of the genes of their individual bearers, he would not have been troubled by this phenomenon. He noted that most other means of defense, such as stings or disagreeable secretions and odors, can be explained by their beneficial effects on individuals in which they are best developed. Because a tasted caterpillar is likely to die, however, it is difficult to understand how tendencies to be more distasteful than average could spread and cause an overall increase in distastefulness.

Fisher next noted that distasteful larvae tend to travel in tightly knit sibling groups. If there is a positive correlation between the degree of distastefulness and the likelihood of avoidance of that group or kind of caterpillars by a bird that had tasted one of them, then a highly distasteful caterpillar, even if it died as a result of being tasted, might enable more of its siblings to survive and reproduce than a less distasteful caterpillar. Since siblings of very distasteful caterpillars have a higher than average likelihood of themselves carrying the genes leading to greater distastefulness, such genes could spread. Fisher even quantified this phenomenon by noting that "the selective potency of the avoidance of brothers will of course be only half as great as if the individual itself were protected; against this is to be set the fact that it applies to the whole of a possibly numerous brood" (p. 178). In other words, full siblings on average have 50 percent of their genes in common, or identical by immediate descent (ibd). This means that each protected full sibling is 50 percent likely to be carrying an allele causing greater distastefulness in a tasted sibling.

This kind of selection leads to nepotism, or the dispensing of benefits to relatives other than offspring or other direct descendants. Darwin (1859), Haldane (1932), and George and Doris Williams (1957) used similar explanations to account for the appearance of sterile castes in the social termites, ants, bees, and wasps. Hamilton (1964), in the first general development and quantification of the theoretical basis of nepotism, concluded that *"the social behaviour of a species evolves in such a way that in each distinct behaviour-evoking situation the individual will seem to value his neighbours' fitness against his own according to the coefficients of* [genetic] *relationship appropriate to that situation."* This is clearly a prediction of enormous consequence for the analysis of social structure in any species in which genetic relationships correlate with stable social relationships, allowing the evolution of patterns of nepotism.

As Hamilton (1964) noted, the parent-offspring relationship is not fundamentally different from that between other kinds of relatives. In other words, the effort exerted by a parent in rearing its offspring is also a component of nepotism. This leads to the following general axiom: In sexually reproducing organisms, reproductive effort (calories expended and risks taken in the effort to reproduce) is evolved to be expended entirely as nepotism, including (1) parenthood, (2) assistance to relatives other than offspring, and (3) mating effort or the effort involved in placing one's gametes in the best possible environment. This is a particularly important statement for the understanding of sociality since it says, in effect, that sexually reproducing organisms literally evolve to be altruists, but altruists of a very special sort whose altruism, whatever form it takes, is ultimately channeled to genetic relatives. Such altruism, by which the phenotype is used to reproduce the genes, may be described as phenotypically (or self-) sacrificing but genotypically selfish (Alexander, 1974). (See table 1 and figure 5.)

According to inclusive-fitness theory, then, we should have evolved to be exceedingly effective nepotists, and we should have *evolved* to be nothing else at all. Of course, this is exactly what Lack's findings in regard to parental care and clutch size, Fisher's theory of sex ratio selection, and Williams' theory of senescence actually meant all the time—that we are programmed to use all our effort, and in fact to use our lives, in reproduction. Hamilton added the realization that, to test the idea that the maximizing of genetic fitness by individuals has been a central theme in the history of a socially complex species, like ourselves, one has to consider not only parent-offspring interactions, but how we distribute social benefits among all of our genetic relatives (West Eberhard, 1975).

INCLUSIVE FITNESS AND GROUP SELECTION

Despite the recent prominence accorded the view that the maximizing of inclusive fitness by helping the aggregate of one's relatives, or what Maynard Smith (1964) called "kin selection," is a kind of group selection (Brown, 1966, 1974, 1975; E. O. Wilson, 1973b, 1975; D. S. Wilson, 1975a, 1975b; Wade, 1976), this is a misleading if not erroneous view (see also West Eberhard, 1976; Maynard Smith, 1976). Group selection thwarts the reproductive interests of individuals when these interests differ from those of the group; kin selection is a way in which individuals further their genetic interests via other individuals who carry at least some of their genes. Kin selection not only may lead to the favoring of more closely over less closely related individuals within the group, but, even when such discrimination is precluded, it is equivalent to group selection only when the reproductive interests of the group and all of its members are identical. Even in the evolution of distastefulness among sibling caterpillars, what is being saved is not the group as such but the genes leading to distastefulness, wherever they may occur. Acts of altruism among related individuals might sometimes be appropriately termed group selection of *genes* that are identical copies of one another, but not of individuals, which are not. As Wade (1976) points out, and all of the above authors imply, to understand why genes such as those for distastefulness in caterpillars might spread we must understand the structure of groups of caterpillars. Similarly, we must understand exactly how populations of genes are arranged in genomes if we are to understand how mortal individuals derived from those genomes can be the vehicles of genic immortality through their effects on genic reproduction (Leigh, 1977; Dawkins, 1977; Alexander and Borgia, 1978).

Some of the confusion in this realm arises, I believe, because the complex and sophisticated models of group selection in the current literature, which attempt to explain when differential extinction of groups can override differential reproduction by individuals, do not justify the old uncritical notion that the behavior of individuals must somehow be naturally directed toward saving or helping the group as a whole. That idea depended on an implicit assumption that conflicts of interest do not occur below the level of the group, population, or species. It is a defunct idea better described as anti-selectionist than as group-selectionist.

It is important to realize this, because many general treatments of human behavior by social scientists assume that individuals are just vehicles for the welfare of the society. A particularly clear example is the book *Social Structure* (1949) by the anthropologist George Peter Murdock. The book is filled with profoundly important data, and I regard it as one of the most important volumes on the behavior of the human species. Nevertheless, essentially every interpretation or conclusion is developed in terms of a function for the whole society. It is remarkable that twenty-two years after the publication of *Social Structure,* and without evidence that he knew of the revolution in evolutionary biology, Murdock renounced his approach there, stating that "culture" and "social system" are "mere epiphenomena," and that human behavior must be studied as the outcome of the interactions of individuals (Murdock, 1972).

Social Reciprocity

One more idea must be discussed before comparing organic evolution directly to cultural evolution. In 1859, Darwin spoke of the "lowly motive" of helping another individual in the expectation of receiving even more assistance in return. Similarly, in writing of human heroism, Fisher (1958, p. 265) noted that it is possible for heroic qualities to be selected "beyond the limits set by prudence, by a method analogous to that used . . . to explain the evolution of distasteful qualities in insect larvae. The mere fact that the prosperity of the group is at stake makes the sacrifice of individual lives occasionally advantageous, though this, I believe, is a minor consideration compared with the enormous advantage conferred by the prestige of the hero upon all his kinsmen."

Fisher again stubbornly seeks explanations which do not invoke the unlikely phenomenon of group selection. He acknowledges that nepotism may explain heroism, then adds another factor—the effect of the prestige of the hero upon his kinsmen. Such prestige, and its accompanying benefits, are necessarily conferred by nonkinsmen. Heroism would thus arise and be maintained at least partly because of an implicit promise or guarantee within human society that anyone displaying unusual heroism in protecting or saving nonrelatives could expect his relatives to benefit even if he were killed in the effort. Laws that prohibit certain acts are also societal guarantees or promises—of punishment, or of rewards for *not* committing certain acts. In both cases the principle is that of reciprocity, very widely considered by social scientists, and first discussed in detail in an evolutionary context by Robert L. Trivers (1971). (For anthropological discussions, mainly of nontechnological or "primitive" societies, see Malinowski, 1926; Durkheim, 1933, 1938; Mauss, 1954; Sahlins, 1965. For sociological and psychological discussions, mainly of modern, urban, or technological societies, see Homans, 1961; Blau, 1965; Emerson, 1969; Simpson, 1972; Berkowitz and Walster, 1976.) Reciprocity can be divided into two types (figure 5). Direct reciprocity occurs when rewards come from the actual recipient of beneficence. Indirect reciprocity, on the other hand, is represented by rewards from society at large, or from other than the actual recipient of beneficence. We engage in both kinds more or less continuously. In the broadest sense reciprocity is illustrated by many kinds of cooperation, the buying and selling of goods,

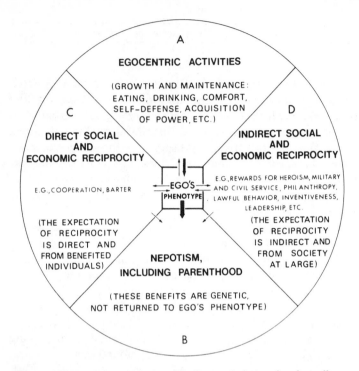

Figure 5. The organism as nepotist. The diagram is designed to show all routes by which expenditures of calories and taking of risks by humans can lead, both directly and indirectly, to genetic reproduction. Reproduction will be maximized when the benefits from egocentric activities and reciprocal transactions maximally exceed their costs, and when the benefits are channeled to the closest relatives with the greatest ability to use the benefits to maximize the reproduction of their relatives in turn. I suggest that as sociality increases in complexity, during evolution, we move from a relatively nonsocial beginning (A), to nepotism (B) expressed only to offspring, and later to nondescendant relatives as well, then direct reciprocity (C) is introduced (chiefly in humans), and finally indirect reciprocity (D), which is probably unique to humans. In human history, expansion of nepotism to nondescendant relatives and increasing group size must have set the stage for an increasing prominence of reciprocal transactions. In large and highly organized states or nations, as compared to simpler forms of social organization, indirect reciprocity must generally be more prominent—indeed, probably a criterion. Simple bands must have been predominantly systems of nepotism to descendant and nondescendant relatives with relatively little complexity in reciprocal transactions. Opportunities to engage advantageously in reciprocal transactions must have begun to appear as systems of nepotism among nondescendant relatives became extensive and complex, which also increased the potential for cheating by recipients of nepotism. Selective pressures leading to larger groups thus promoted increased engagement in reciprocal transactions, and increasingly elaborate social cheating and ability to detect and thwart it. (From Alexander, 1977b.)

and all forms of trading, barter, or transactions. Some reciprocal transactions may involve long delays, and because important risks are covered they may never require actual reciprocation beyond a promise or guarantee; insurance policies and all kinds of written rules or codes are examples.

In nepotistic interactions the reward to the altruist is genetic; in reciprocal systems it is measured in terms of altruism returned to one's self or one's genetic relatives. In nepotism the net cost or benefit of a given act depends upon three variables: (1) the genetic relatedness of altruist and beneficiary, (2) the ability of the recipient to convert the benefits into reproduction, and (3) the cost of the act to the altruist, as measured by available or likely alternatives. In reciprocity there is only the cost of the act to consider and the likelihood of returns of a greater value. One way to increase the likelihood of reciprocity in altruism is to restrict the altruism to situations in which repayment is immediate, as in social cooperation to thwart a mutual threat, or the simultaneous exchange of items of value. Such transactions ought to typify brief interactions between parties relatively unknown to one another or unlikely to interact thereafter, since such individuals will either be ignorant of the likelihood of delayed reciprocation or incapable of applying pressure should it not be forthcoming. When reciprocal exchanges are more or less continuous and essential, and involve more than two individuals, it is also possible to lose by reneging on a debt because of the response of observers with whom one will need to interact later. This means that reciprocal interactions should involve greater commitments or apparent risks when they occur in the presence of witnesses who may subsequently bring social pressure to bear on any who fail to fulfill their part of the bargain. It is particularly important to understand reciprocity not only because it applies generally to male-female interactions, but also because, as I shall argue later, it seems to be characteristic of humans everywhere, and it has become the essential social cement in all large modern nations (Alexander, 1978d).

Historical Relationships between Nepotism and Reciprocity

Although evolutionists tend to regard nepotism and reciprocity as entirely distinct, the probable historical and functional relationships between them are relevant to a biological perspective on human affairs. Alexander and Borgia (1978) argue that nepotism to nondescendant and distant descendant relatives is an extension of (evolutionarily) earlier altruism in the form of parental care, citing the ubiquity of parental care in social organisms, the higher accuracy in assessing parenthood as compared to other relationships, and the paucity of evidence for undirected altruism toward all, as opposed to altruism directed toward individual relatives of different degree. I have argued (Alexander, 1979c) that reciprocity is in turn largely derived from nepotism.

Pure social reciprocity can be defined as interactions that normally occur between nonrelatives, the structure and circumstances of which indicate clearly that each party may be expected (by the observer) or expects (in the case of humans) to receive benefits greater than those given. This outcome is possible whenever there is an asymmetry between the respective needs of the participants and their respective abilities to give. For example, in a simple sale, the buyer is willing because he has money and needs the product, while the seller is willing because he has the product and needs money. Such exchanges are easy to analyze without an evolutionary perspective because the items of exchange and their values are apparent (e.g., see Homans, 1961, p. 62, for a nonevolutionary analysis using phrases almost identical to those used here).

Except for male-female interactions (Alexander and Borgia, 1978) pure reciprocity (that is, exchanges between nonrelatives that involve temporarily costly acts of altruism by each participant) has probably never been demonstrated for a nonhuman species, and it may actually occur only infrequently in humans, outside modern nations. Although Trivers (1971) has argued for nonhuman cases of reciprocity involving cleaning and cleaned fish of different species, and alarm calls in flocks of birds, both cases may be open to doubt. West-Eberhard (1975) and Sherman (1977) have explained the various ways in which alarm calls can be viewed in strictly selfish or nepotistic terms (see also the earlier discussion of altruism), and there is no evidence that either fish species in the cleaner-cleanee relationship ever actually behaves altruistically. The cleaner may be simply acquiring food in the best way available to it, and the cleaned fish may refrain from consuming its cleaner not because the cleaner will then be available for subsequent cleaning but, in evolutionary terms, because the cleaner is poisonous and ill-tasting. Neither fish, then, would actually be engaging in an altruistic act calling for reciprocity.

The reason for suggesting uncommonness of pure reciprocity in humans other than in modern nations is that in the small bands in which humans are generally presumed to have lived throughout most of their evolutionary history, practically all social interactions were among relatives. Thus, Wiessner (1977) found that !Kung Bushmen almost never engage in "exchange" (xharo) with individuals not known to be related to them. Among Yanomamö Indians, neighboring tribes, and even enemy groups, are nearly always composed of known relatives (Chagnon, 1979a,b, and personal communication).

The concept of "pure" reciprocity returns us to the question of the relationship of nepotism to the origins of "pure" reciprocity. Although either nepotism or reciprocity can occur in the absence of the other, the two probably most often occur together. Thus, every act of nepotism (as with any aspect of reproductive effort) involves some risk. Even a mother nursing a baby and, for purposes of this example, expecting no altruistic returns at all (hence, engaged in "pure" nepotism) would be, like a bank loaning money (hence, engaged in "pure" reciprocity), making an investment which carries with it a certain risk. The baby may die of a childhood disease, it may be sterile, or for some other reason it may yield no genetic returns whatever to the mother. (Note: In an example like this I am not suggesting anything at all about the conscious motivations of the mother, or her attitude toward her own motivations, although the intense emotional concern and satisfaction of mothers in this kind of circumstance can scarcely be regarded as irrelevant to a history of differential reproduction.)

"Pure" nepotism, then, is represented by cases in which an organism that gives benefits to a genetic relative expects absolutely nothing in return except an increased reproduction of the genes it shares with the recipient of its altruism. Such cases must be less numerous than is commonly suspected, although much of parental care to very young and helpless juveniles would qualify, and certainly so in species in which the parent tends offspring briefly, then leaves or dies and never again interacts with them.

Nepotism in humans, however, usually takes a somewhat different form. Thus, anyone who helps a cousin or other distant relative may be satisfied even if no direct compensation ever occurs; on the other hand he is likely to expect that the relative will return the favor. Even parents, among humans, often behave as though they expect their offspring to return certain kinds of favors or assistance to them because they have provided for the offspring (see also Irons, 1979a). I suggest that this tendency relates to the long period of life during which humans are able to reproduce, or redistribute resources in their own interests, a period that commonly overlaps the period during which their offspring become able to redistribute resources, and during which the parents' interests frequently differ from those of their offspring (Trivers, 1974; Alexander, 1974). That exchanges between relatives lead to special kinds of expectations is suggested by popular wisdom such as the often-heard admonition to be cautious about "doing business" with a relative.

Reflection upon cases in which nepotism and reciprocity are mixed, then, shows that the differences between them are not so profound as they might at first seem. Each is an investment with some risk. This means that, from a reproductive or evolutionary viewpoint if a social investment has some likelihood of not being returned it is better given to a close relative than to a distant one. Then, if reciprocity fails, there will at least be a chance of genetic return. It is likely that, during human history, disparities in the ability to give and the ability to use continually arose among social interactants (who in primitive societies would essentially always be relatives); these conditions would encourage social investments of a sort ordinarily calling for reciprocity but not necessarily requiring it for the interaction to have at least some reproductive benefit for the investor. Even in pure reciprocity the eventual destination of any resource garnered is, again in historical terms, redistribution via nepotism in the interests of genetic reproduction. The adage about "doing business with a relative" is best translated to mean that doing *business* with genetic relatives may severely prejudice the likelihood of overcompensating return—in other words, that relatives are more likely to feel free to renege on a debt, and to get away with it should they try. In-laws, who are often included among one's relatives (the ways in which this comes about, and the reasons why we accept it, are themselves interesting to contemplate), are notoriously regarded as poor risks in transactions. We need only reflect that we have no genetic interest whatever in our spouse's relatives (unless the spouse is a relative), hence, we always suffer a nepotistic loss if altruism is not returned from that "side of the family." Thus, married adults (at least those who have never thought about the arguments presented here) might be expected to look favorably upon investments in their spouse's side of the family only if the spouse's family is wealthy, or for other reasons quite likely to more than repay the debt.

I am not suggesting that we cannot behave contrarily to such expectations from evolutionary theory, even to the extent of neither having children nor helping relatives at all. But I do suppose that situations in which we might learn to do such nonreproductive things have been quite rare during human history. Nor am I suggesting that all of the above "expectations," or even any of them, are necessarily conscious, only that we may be expected to behave appropriately to them.

Without an understanding of the centrality of genetic reproduction, and the concept of inclusive fitness, it has been impossible to develop a general theory of sociality from the analysis of social reciprocity. The interactions that have most puzzled social scientists in this effort have been those involving chiefly one-way flows of benefits, and commonly taking the form of "deep" or "intimate" relations (e.g., Homans, 1961; Sahlins, 1965; Hatfield et al., 1979; Alexander, 1975a, 1979c; figure 6). Of course, these interactions are the very ones most likely to involve genetic returns and least likely to require actual reciprocity, hence most confusing to anyone not considering the significance of our history of natural selection.

It seems clear that *if we are to carry out any grand analysis of human social behavior, it will have to be in evolutionary terms, and we shall have to focus our attention almost entirely upon the precise manner in which both nepotistic and reciprocal transactions are conducted in the usual environments in which humans have evolved their social patterns.* That it has taken us so long to find this out is only a consequence of our never having been endowed by our genes with the capacity to develop, in our usual environments, sensory abilities enabling us to detect the genes themselves and to understand their mission; those abilities came incidentally with science and technology, themselves incidental outgrowths of the nepotistic striving of the successions of our ancestors.

So it is not surprising that Darwin could not answer the question: survival of the fittest *what?* For the *what* is genes and groups of genes about which he knew nothing. Despite its philosophical impact, Darwinian evolution did not initially provide the solution to the age-old conundrum of whether humans really are hedonistic individualists or group altruists. Only knowledge of genes could tell us that, in fact, they are both and yet neither. They seem to be individualists because each individual tends to behave according to his separate genetic interests, but each individual actually and literally uses or gives his life for others, who carry the copies of his genes. For one hundred years after Darwin, biologists floundered over choosing, for organisms in general, between the same two anciently opposed views of humans.

Now we can proceed from the knowledge that the unity of

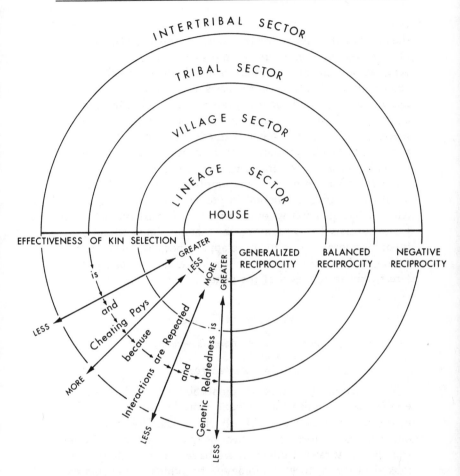

Figure 6. This diagram from Sahlins (1965) illustrates the different types of what he called "social reciprocity" in primitive cultures. The information in the lower left quadrant has been added to suggest how kin selection and evolutionary principles accord with reciprocity, as practiced by human groups. In effect, "generalized reciprocity" involves mostly one-way flows of benefits because it is largely nepotism (the return is genetic), and "negative reciprocity" involves one-way flows because it consists of one-time interactions accompanied by a great deal of social cheating. "Balanced reciprocity," on the other hand, tends to occur between distant relatives or nonrelatives that are likely to interact repeatedly, and therefore involves balanced flows of benefits. Studies made by "exchange" and "equity" theorists in psychology and sociology focus on what in Sahlins' terminology would be balanced reciprocity. (From Alexander, 1975a.)

the individual derives from the ability of the tens of thousands of separately heritable genetic units making up its genotype to act in all of their separate interests equally by dispensing altruism to other individuals on the basis of the *probability* that such individuals carry any particular one of the genes of the altruist: on the probability that it is an offspring, a sibling, a nephew or niece, a cousin, or some other distinctive class of relative, carrying some particular proportion of the genes of a potential altruist. This I regard as a general theory of the individual, and as a theory of human nature. It is simple to state, and it resolves all of the paradoxes inherent in previous theories—most specifically the question whether we are selfish individualists or group altruists. For the same reason that it is simple to state, this theory is enormously difficult to apply—because, as I have said earlier, it drives the question of understanding conflicts of interest right to the level of the gene.

WHY LIVE IN GROUPS?

*Nowhere on earth do people live regularly
in isolated families.*
—*George Peter Murdock, 1949, p. 79*

To the extent that these arguments about individuals having evolved to maximize their genetic reproduction apply to humans, the consequences are clearly enormous. They must affect our views of almost every aspect of human behavior. As one example they raise a question we have almost entirely overlooked in the history of biology: what kinds of benefits can cause group-living to be profitable to individuals, and what are the consequences for individual behavior if group-living is mandatory? What if living in a particular group already having a particular structure and a particular set of social rules is mandatory, as it is for most of us today?

Sociality by definition can exist only in organisms that live in groups. Efforts to understand sociality for a long time rested upon the view, taken for granted, that groups exist for the good of the species, and individuals for the good of the group. If groups form, and the individuals within them interact and co-operate, solely to help perpetuate the species, then deleterious consequences to individuals are to be expected and will not necessarily be minimized. Only the success of the group matters. As soon as we suppose that organisms do things because, in historical terms, they thereby help their own personal reproduction, questions are raised about all of the events ordinarily regarded as "social," such as cooperation, sharing, and all forms of altruism. Our attitude toward voluntary group-living is also changed. If behavior evolves because it helps individuals, then our attention is drawn to the fact that the crowding or aggregation of group-living inevitably entails expenses to individuals, such as increased competition for all resources, including mates, and increased likelihood of disease and parasite transmission (Alexander, 1974). Accordingly, for the first time really, we are caused to wonder why animals should bother to live in groups. Why be social, beyond what is required to mate and raise a family? If the answer is that individuals living in groups reproduce more than individuals not living in groups, then, also for the first time, we are led to seek out the specific reproductive benefits that accrue from social life. Except in clones (groups of genetically identical, asexually produced individuals), the interests of individuals within groups are never identical with those of the group as a whole, and a basic problem in understanding sociality is to specify the precise kinds and amounts of conflict of interests among individuals within groups, and their results.

The automatic detriments of group-living can be understood only by realizing the separate interests of the individuals involved. Consider the subordinate male of a baboon or other primate species rendered effectively sterile by an aggressive dominant who keeps him away from the ovulating females, or the dominant male who is cuckolded by the sneaky subordinate he has not ostracized completely or killed. Consider the female of a colony-nesting bird or a polygynous primate unable to secure all the parental attention of the father of her offspring because of other females nearby. Consider those gulls, swallows, penguins, or anis who, for reasons not always clear to us, must nest very close to one another where there is maximal risk of having another bird deposit its eggs in their nests. Consider the herbivore or carnivore who must continually tolerate other nearby individuals simultaneously seeking the best food or the safest feeding locations.

For reasons such as these, group-living must be an attribute like extended juvenile life and lowered clutch or litter sizes; in each case the attribute has evolved only because benefits specific to the organism and the situation have outweighed what apparently are automatic detriments. Longer juvenile life and lowered clutch or litter size seem to lower reproductive rates (i.e., the rate at which individuals' genes are replicated), but of course they only lower *potential* reproductive rates that may never be approached in the natural environment. The benefits of lengthened juvenile life, which counterbalance any detriments, may be greater adult size, increased time for learning critical to survival or reproduction, better timing of resistant stages with harsh seasons, or conservation of reproductive energy and risk-taking until some optimal time. The benefit of lowered clutch or litter sizes is to maximize genetic representation at some subsequent time—say at fledging, weaning, or breeding time; one produces fewer offspring so as to rear them better and produce more grandchildren. We are left with the question, then, of what are the benefits of group-living which offset its automatic detriments?

Theoretically, the causes of group-living include any enhanced access to some resource, or enhanced ability to exploit some resource, which is sufficient to more than offset the automatic detriments described above. In fact, though, an exhaustive list of the selective backgrounds favoring group-living in all organisms, may contain no more than three general items (Alexander, 1971, 1974, 1975b): (1) susceptibility to predation (by members of one's own or some other species) may be lowered either because of aggressive group defense, as has been reported in musk ox (Tener, 1965), or because of the opportunity for individuals to use the group as cover (or to cause other individuals to be more available to predators), as with schooling

fish and herds of ungulates (Hamilton, 1971) and the members
of Batesian and Mullerian mimicry groups (Wickler, 1968); (2)
the nature of food sources may make it unprofitable for individuals to splinter off, as with (a) wolves dependent upon
large game in certain regions (Mech, 1970), or (b) groups dependent upon scattered large supplies of food that individuals
locate too infrequently on their own (as perhaps with vultures);
or (3) some resources may be extremely localized, such as cliffs
where hamadryas baboons can sleep secure from predators
(Kummer, 1968), or predator-free islands and cliffs where ma-
rine birds and mammals can breed (e.g., Ashmole, 1963; Bartholomew, 1952).

The asymmetry of these three items suggests the difficulty of
attempting precise definitions of "social groups" or "group-
living" (see also Horn, 1971; Lack, 1968). In the first two cases
the individual who is in a group gains because of the presence
of the other individuals; in the third it does not, but instead
gains solely from the presence of some other resource in the
immediate environment (that is, other sources of mortality do
not keep the population low enough to prevent extreme competition for the localized resource). In the first two cases, then, one
expects individuals to approach or remain near other individuals. In the third case individuals may aggregate around resources but are otherwise expected to avoid one another, or to
be aggressive, although they may use the presence of other
individuals or aggregations as indicators of resource bonanzas.

I suggest that group-living in any organism, including humans, only appears because one or some combination of these
three general extrinsic causative factors renders individuals accepting the automatic detriments of group living more fit than
solitary individuals (figure 7).

On the other hand, some unusual cases may not fit well into
the three general categories described above. Thus, communal
clusters (e.g., of flying squirrels) may chiefly gain from mini-
mizing energy loss (Muul, 1968), and in the V-formation of
migrating waterfowl individuals may gain from pooling their
information about the long migratory route (W. J. Hamilton,

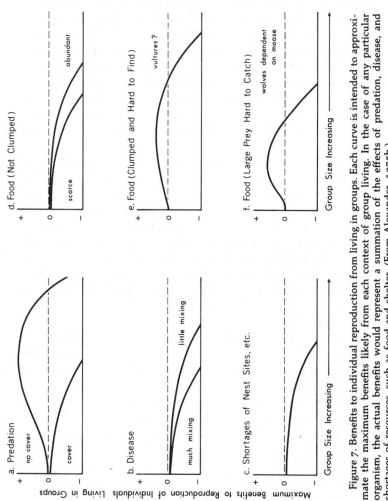

Figure 7. Benefits to individual reproduction from living in groups. Each curve is intended to approximate the maximum benefits likely from each context of group living. In the case of any particular organism, the actual benefits would represent a summation of the effects of predation, disease, and shortages of resources, such as food and shelter. (From Alexander, 1975b.)

1969). Males in non-resource-based leks may aggregate and display because females have evolved to mate only when numerous males can be compared (Alexander, 1975b). There may be other resources less obvious than protection from predators or increased access to food that are made more available by group cooperative effects. None, however, is obvious to me as an actual cause of group-living or its maintenance. Freeland (1976) argues that disease may be another primary cause of group-living, the stability of the group reducing the likelihood of disease by reducing the flow of individuals between groups (especially venereal and other socially transmitted diseases). But his argument seems to me only to account for modifications of group-living, once other causes establish and maintain it, because of the potential for increased disease transmission under group-living. Similarly, the argument that group-living may be initiated in cooperative defense of clumped resources (e.g., Brown and Orians, 1970) seems to me to refer to a secondary effect following simple grouping (i.e., without group defense) on clumped resources.

The view just outlined differs markedly from those prevalent during most of the history of the study of sociality. The general opinion for long had been that group-living and cooperativeness are universally and automatically beneficial to all concerned (and indeed that on this account they are basic attributes of all life). The roots of this opinion are found in opposition to application of the "nature red in tooth and claw" interpretations of Darwinism to include social behavior (Hofstadter, 1955; see also references in Allee, 1932; Wheeler, 1923). This opposition was reinforced by a succession of influential writers such as Kropotkin (1902), Allee (1951), Montagu (1955), and Wynne-Edwards (1962). Until a few years ago, the notion that evolution leads largely to acts that are group-beneficial was assumed to be correct by nearly all social biologists. I cannot avoid the impression that the tenacity of this view is partly due to the way humans think about their own lives and purposes. We seem to do more or less as we please, and we think of ourselves as cooperative and altruistic.

Ironically, the argument that humans are basically cooperative and altruistic is just as instinctivist or deterministic as its
counterpart that they are basically aggressive and competitive.
Nonetheless, and however inappropriately, opponents of the
latter view are usually thought of as staunch anti-instinctivists,
whatever the implications of their views about social behavior.
The effect of our views about ourselves upon our assessments
of social behavior becomes an unexpectedly intricate problem.
I mention it here because the logical consequence of accepting
an extreme "basic social instinct" or "innate social appetite"
view is that group living, cooperation, and altruism require no
special explanation. In the opposing view, just espoused, they
do, and the number of alternatives is small.

To my knowledge, no reasons for group-living have been
proposed, other than those given above, and consistent with the
now well-established fact that selection operates principally on
the genetic replicators possessed by individuals. This is a crucial
point. It leads to the expectation that individuals should avoid
competitors and be nonsocial, and that large groups should
appear only (i) when the resources of reproduction are so
clumped that individuals must come into close proximity (item
3 above), a situation implying no cooperation, hence no special
social organization, or (ii) when cooperation contributes to individual reproduction in the population or species at large because of some hostile environmental force (items 1 and 2
above).

Applying these hypotheses to familiar organisms yields some
interesting and surprising results. Thus, item 2a applies only to
species like African hunting dogs, wolves, lions, group-fishing
pelicans, and group-hunting fish; and item 2b applies easily to
only a few animals, like foraging vultures and sandpipers. This
leaves item 1—predation, from members of one's own or another species—as the probable selective background of most
other actively formed groups, including those of most primate
species and all of the great herds of ungulates and schools of
fishes. I believe that this conclusion is supported by intensive
studies of group-living conducted since I first (Alexander, 1971,

1974) proposed these arguments (e.g., Hoogland and Sherman, 1976; Hoogland, in press, a-c).

In summary, biologists have recently generated a new and apparently more precise picture of the evolutionary process. With natural selection as its principal guiding force, evolution has led to the production of individuals, created by temporary coalitions of chromosomes, which are themselves temporary coalitions of genes. These individuals first gather resources as they grow, develop, and learn, and then dispense those resources in their own genetic interests by all available means, including, for species like ourselves, the production and assistance of offspring, nepotism to needy nondescendant relatives, and exchange behavior of various sorts with both relatives and nonrelatives (figures 4–6). Even when humans live in variously cooperative and socially complex groups they do so because, historically, group-living has enhanced the reproductive success of individuals.

This general view of organic evolution I regard as firmly established; I do not expect any significant part of it to be retracted or altered in the future. As Williams (1966) suggested, this view gives to natural selection an enhanced explanatory potential that essentially eliminates the reasons for invoking a host of questionable concepts prevalent during the history of evolutionary study. Examples are the ideas of "progress" and "orthogenesis" in long-term evolution, the tendency to invoke genetic drift or mutation rates to explain long-term trends, and the implication that phenomena like allometry (structures changing in size together) and neoteny (appearance of juvenile attributes in adults and vice versa) occur in spite of natural selection rather than because of it. With respect to our topic here, the new view of natural selection provides a solid theoretical base from biology with which to pursue the questions of human sociality and the nature of culture.

2

Natural Selection and Culture

By culture *we mean an extrasomatic, temporal continuum of things and events dependent upon symboling. Specifically and concretely, culture consists of tools, implements, utensils, clothing, ornaments, customs, institutions, beliefs, rituals, games, works of art, language, etc. All peoples in all times and places have possessed culture. . . .*
                                   —White, *1959, p. 3*

Thus, we are brought to what many regard as the most distinctive attribute of our species—its possession of culture. What are the relationships between the reproductive success of individuals and the existence and nature of culture? Is it possible that once the capacity for culture was achieved the actual directions of cultural change and the patterns of culture in different societies simply became arbitrary in relation to the reproduction of genes of individuals? Culture is heritable, but through learning, not directly through the genes. Culture changes, but not by genetic mutations, and sometimes massive changes can occur without genetic changes. Most social scientists regard culture as adaptive, but in biological terms their interpretations seem paradoxical. Culture, after all, appears to be a group phenomenon, both because it extends unchanged, in many cases across multiple generations and far beyond the individual's lifetime, and because the extreme complexity and division of labor in modern societies make the individual so completely dependent upon the rest of society. For these reasons the biological significance of culture has remained an unsolved problem. In the light of the revolution in evolutionary biology, the question has now become: to what extent can culture be understood by thinking in terms of individuals striving to maximize their genetic reproduction?

Probably because anthropologists and others have tended to identify the possession of culture and the capacity for culture as peculiarly human traits, the concept of culture has acquired and retained a certain singularity. One result has been a proliferation of general theories about the function or value of culture. No such theory seems to have acquired or maintained wide acceptability. This failure has caused some investigators reluctant to see culture as a generalizable phenomenon to assert that "Culture is dead"—that it does not possess the singularity attributed to it and that truly general theories of culture are therefore unlikely. Rather than the search for a general theory being abandoned completely, however, the idea has developed in some circles that, in the absence of acceptable functional explanations, culture can only be explained in terms of itself, or as a set of arbitrarily assigned meanings or symbolizations (White, 1949; Sahlins, 1976b), and specifically cannot be explained in terms of utilitarianism of any sort or at any level. White (1949) promoted an extreme version of this argument when he wrote of how fallacious is the fond belief that it "lies within man's power . . . to chart his course as he pleases, to mold civilization to his desires and needs" (p. 330).

Such scientists argue that culture is something greater than the activities of humans collectively and almost independent of the activities of humans as individuals: it continues on its own course, which is probably unpredictable and certainly swayed only slightly by the wishes of individuals, who are merely its "passive" transmitters. A recent version of this argument was developed by Marshall Sahlins in his book *Culture and Practical Reason* (1976b). There he comes very close to describing culture as an aspect of the human environment about which we can do little but accept it in just those terms. This view of culture parallels the biologists' concept of the genotype and the phenotype as parts of the environment of selection of the individual genes.

If human evolution, like that of other organisms, has significantly involved selection effective at genic levels, realized through the reproductive strivings of individuals, neither humans as individuals nor the human species as a whole have had a single course to chart in the development of culture but rather a very large number of slightly different and potentially conflicting courses. In this case it would indeed be difficult to locate "a function for," or even "the functions of," culture. Instead, culture would chiefly be, as Sahlins' view may be slightly modified to mean, the central aspect of the environment into which every person is born and where one must succeed or fail; developed gradually by the collections of humans that have preceded us in history; and with an inertia refractory to the wishes of individuals, and even of small and large groups. Culture would represent the cumulative effects of what Hamilton (1964) called inclusive-fitness-maximizing behavior (i.e., reproductive maximization via all socially available descendant and nondescendant relatives) by all humans who have lived. I regard this as a reasonable theory to explain the existence and nature of culture, and the rates and directions of its change.

If this theory is correct, then each of us should find some aspects of culture objectionable. Seldom would any of its aspects be viewed with equal good humor by all of us, and in just this circumstance we would not expect easy success for grand utilitarian views of culture, general theories of culture, or efforts at deliberate guidance of societal change. These are exactly the kinds of failures that have always plagued culture theorists.

CULTURAL INERTIA

The inertia of culture has suggested to some anthropologists that individuals—even great persons—have scarcely any impact upon its nature and its directions or rates of change. I suggest that this is not because culture is nonfunctional or mysterious but because the underlying pattern of conflicting interests has been too complex and too poorly understood to be grasped within a single framework. Yet, by the theory developed here, the inertia of culture would exist *because* individuals and subgroups had influenced its direction and shape, molding it—even if imperceptibly across short time periods—to suit their particular needs, thereby incidentally increasing the likelihood that subsequent individuals and subgroups could find ways to use it to their own advantages as well, and could not alter it so greatly or rapidly. Some confusion may derive from the fact that reproductive efforts by individuals would not necessarily be directed at *changing* culture, nor would such efforts lead to any particular directions of change in culture as a whole. Instead, the striving of individuals would be to *use* culture, whether or not change were entailed, to further their own reproduction. No necessary correlation would be expected between the success of an individual's striving and the magnitude of the individual's effect on cultural change, or between the collective success of the individuals making up a group or society and the rate of cultural change. It would not matter if one were a legislator making laws, a judge interpreting them, a policeman enforcing them, a lawyer using them, a citizen obeying or circumventing them, or a criminal violating them: each of these behaviors can be seen as a particular strategy within a society governed by law, and each has some possibility of success. They are efforts, by and large, to exploit rather than change culture. It would tend to be contrary to the interests of the members of society that cultural changes of any magnitude could easily be effected by individuals, except for inventions seen as likely to benefit nearly everyone. The reasons are that (1) changes, effected by individuals or subgroups in their own interests, would probably be contrary to the interests of others, and (2) once individuals had selected and were carrying out a particular set of responses to the existing culture in their own interests, changes of almost any sort would have some likelihood of being deleterious to them. These arguments not only suggest how anthropological interpretations of culture may be entirely compatible with the notion of reproductive striving principally effective at the individual (or genic) level, but they may also explain in part the genesis of the peculiar views from within some branches of anthropology that culture is somehow independent of individuals and groups and their wishes, and not easily explainable in any kind of utilitarian terms.

If the "rational self-interest" of "utilitarian" explanations of human behavior is really the combination of the interests of genes and chromosomes in perpetuating themselves—that is, if it is a consequence of the differential survival of genetic replicators across all of history—then predicting or identifying the outcomes of a utilitarian theory of culture is going to be an absolutely stupendous task. We will have to identify conflicts and confluences of interest at every level of human social organization, compromises and coalitions of every imaginable variety and outcome, and power differentials and all of their diverse effects. At every stage of analysis we will have to take into account the consequences of humans having devised cultural items and traits and institutions that persist across generations, and are variously stable and manipulable by all of those self-interest-seeking individuals and coalitions. Finally, our task will be made immensely more difficult by the uniquely extensive parental and nepotistic behavior of humans, by their learning flexibility, their massive alterations of their environment, and the complex mixing and overlapping of their generations. The realization of these complexities makes it difficult for me to agree with those who would term the biological utilitarian approach to cultural analysis an oversimplifying or "reductionist" argument.

Cultural Change

It is a fundamental characteristic of culture that, despite its essentially conservative nature, it does change over time and from place to place. Herein it differs strikingly from the social behavior of animals other than man. Among ants, for example, colonies of the same species differ little in behavior from one another and even, so far as we can judge from specimens embedded in amber, from their ancestors of fifty million years ago. In less than one million years man, by contrast, has advanced from the rawest savagery to civilization and has proliferated at least three thousand distinctive cultures.
—George Peter Murdock, 1960, p. 247

Having stated a general hypothesis about culture, we may now return to questions raised earlier about cultural change. If long-term changes in human phenomena, as evidenced for example in the archaeological record, are cultural, and were not induced by natural selection or accompanied by genetic changes relating to cultural behavior, then what has guided cultural evolution? What degrees and kinds of correspondence exist today between the patterns of culture and the maximization of genetic reproduction of the individuals using, transmitting, and modifying culture? Are the existing degrees and kinds of correspondence, and of failure to correspond, consistent with the forces presumed to underlie rates and directions of cultural change?

At one end of a spectrum lies the possibility that all cultural changes during human history have been utterly independent of genetic change, neither causing it nor caused by it. At the other end is the possibility that changes in human behavior have correlated with genetic change to approximately the same degree as changes in the behavior of other species, such as nonhuman primates. Observations within recorded history are sufficient to show that neither of these extreme possibilities is likely. As an example denying the first possibility, cultural changes, such as inoculations for potentially fatal childhood diseases, obviously influence genetic change; and for the second case, cultural changes clearly have accelerated tremendously in recent decades without evidence of a parallel acceleration in genetic change. At least, then, cultural changes can cause genetic change, although there is no clear evidence that genetic changes (such as might affect behavioral plasticity or speed of learning) are causing cultural changes, or that any close correlation exists between genetic changes and cultural changes that involve behavior. Since it is easy to understand, on theoretical grounds, how culture can change cumulatively without accompanying genetic changes that pertain to the relevant behaviors —and since it is easy to point out that such changes have occurred frequently within recorded history when strikingly different cultures merged—the significance of the above questions about (1) the forces which change culture, and (2) the relationships of culture to maximization of reproduction by individuals, is brought into an even sharper focus. We expect that the answers to these questions will be complementary, and that the efforts to answer them should be conducted simultaneously.

Some changes in culture, such as those caused by climatic shifts, natural disasters, and diseases, predators, and parasites (of humans and the plants and animals on which they depend), are difficult for humans to control or even beyond control; others may be controlled directly and deliberately (as with inventions and conscious planning) or caused almost inadvertently (as with depletion of resources and increases in pollution). The difficult question, in understanding the relationship between culture and natural selection, is not in discovering the reasons behind cultural changes, as such, which are actually fairly obvious (see below). Instead, it is in understanding exactly how such phenomena influence culture: What is done with them? What directions of change do they induce, and why? Those changes in culture resulting from human action appear to represent products of the striving of both individuals and groups. Such changes, as with extrinsically caused changes, are also *responded to* by changes in the striving of individuals and groups. Inventions are seized upon; pollution and resource depletion are lamented, and cause geographic shifts in population or efforts either to offset or take advantage of their effects. Attempts are made to predict and offset natural disasters and climatic shifts. All of these responses are easily interpretable as efforts by individuals, acting alone or in groups, to use culture to their own advantage. But culture is not easily explainable as the outcome of striving to better the future for *everyone equally:* if that were the case, then conscious planning would quickly become the principal basis for cultural change, and it would be carried out with a minimum of disagreement and bickering.

A COMPARISON OF ORGANIC AND CULTURAL EVOLUTION

It is possible to examine the problem of cultural change in a fashion parallel to that used earlier to analyze evolutionary change. We can consider the effects of the same five processes that characterize genetic change (the most closely parallel argument is probably that of Murdock, 1960).

Inheritance: Just as the morphological, physiological, and behavioral traits of organisms are heritable through consistency in the developmental environment, the traits of culture are also heritable through learning. They may be imitated, plagiarized, or taught.

Mutation: Like the genetic materials, culture is also mutable, through mistakes, but also through discoveries, inventions, or deliberate planning (Murdock's "variations," "inventions," and "tentations").

Selection: As with the phenotypic traits of organisms, some traits of culture in some fashion, by their effects, reinforce their own persistence and spread; others do not, and eventually disappear for that reason (Murdock's "social acceptance," "integration," and "selective elimination"; see also Campbell, 1965).

Drift: As with genetic units, traits of culture can also be lost by accident or "sampling error." The last stonemason may die without passing on his knowledge, or the ship full of dissident emigrants may sail to found a new society without anyone realizing there is no stonemason on board.

Isolation: As with populations of other kinds of organisms, different human societies become separated by extrinsic and intrinsic barriers; they diverge, and they may contact and re-merge or continue to drift apart; items and aspects of culture may spread by diffusion (Murdock's "cultural diffusion" and "cultural borrowing").

Immediately, differences are apparent between the processes of change during genetic and cultural evolution. Perhaps most profoundly, causes of mutation and selection in cultural evolution, unlike those in genetic evolution, are not independent. Most of the sources of cultural "mutations" are at least potentially related to the reasons for their survival or failure. A new kind of plow or computer is built for its expected usefulness; it is adopted if it works.

Some culture theorists have nevertheless denied utilitarian connections between the usual sources or causes of cultural change and the reasons for their survival or failure. I suggest that one reason for these denials is that such theorists have never sought function in terms of reproduction at the individual level, as biologists now realize must be the case in organic evolution. Some, such as Franz Boas (1911, 1940), Ruth Benedict (1934), Frederik Barth (1967), and eventually George Peter Murdock (1972), have emphasized the role of the individual in culture, and some, such as Bronislaw Malinowski (1926, 1944), A. R. Radcliffe-Brown (1922, 1952), and Roy A. Rappaport (1968), have emphasized function as survival value—usually for the group or population but sometimes for the individual. Still others, with a psychological background, have argued that responses to the cultural environment by individuals are best analyzed as efforts at "drive reduction" (e.g., Spiro, 1961). None of the culture theorists, however, has seen function as reproductive value. Instead, most functionalists have either sought group-level utilitarian effects or have regarded as crucial the survival of the individual, but not its genetic reproduction (for reviews of the views of these and other anthropological theorists, see Hatch, 1973; Harris, 1971; Sahlins, 1976b).

Among current authors, Sahlins (1976b) perhaps best exemplifies the rejection of functional explanations of culture, denying both individual-level and group-level utilitarianism:

> For some [what he would call the "subjective utility" theorists] . . . culture is precipitated from the rational activity of individuals pursuing their own best interests. This is 'utilitarianism' proper; its logic is the maximization of means-ends relations. The objective utility theories are naturalistic or ecological; for them, the determinant material wisdom substantialized in cultural form is the survival of the human population or the given social order. The precise logic is adaptive advantage, or maintenance of the system within natural limits of viability. As opposed to all these genera and species of practical reason, this book poses a reason of another kind, the symbolic or meaningful. . . . It . . . takes as the decisive quality of culture . . . not that [it] must conform to material constraints but that it does so according to a definite symbolic scheme which is never the only one possible. Hence it is culture which constitutes utility. [Pp. vii–viii]

Malinowski's (1944) "functional" theory of culture, although in striking contrast to approaches like that of Sahlins, was equally unsatisfactory. Because it was couched in terms of satisfying immediate physiological needs, as with learning theory and other theories that stop with proximate mechanisms, it could not account for the existence of those needs (Alexander, 1977b), any more than Sahlins' theory can account for either the existence or nature of symbols. Sahlins (1976b) was led to the conclusion that for Malinowski culture represented a "gigantic metaphorical extension of the digestive system." Malinowski's theory, though, would have made sense if only he had seen culture as a gigantic metaphorical extension of the reproductive system.

Recent investigators of the relationship between organic evolution and cultural change, such as Cloak (1975), Dawkins (1976), Durham (1976b, 1979), Cavalli-Sforza and Feldman

(1973), and Richerson and Boyd (1978, in press) have argued that the separate mode of inheritance of cultural traits is the central feature of culture, and that the significance of this feature is its thwarting of the natural selection of genetic alternatives pertaining to human behavior. I regard this approach to the history of culture as similar to a view of the natural history of organisms that sees phenotypes in general (as opposed to no phenotypes) functioning as thwarters of natural selection. In one sense they do "thwart" the forces of selection. By providing the organism with modifiable responses to varying environmental contingencies, they necessarily render the action of selection on the genes less direct: selection must now act through the phenotype. But this effect is itself a product of selection; it appeared because those genes that reproduced via phenotypes outsurvived their alternatives in the environments of history. So it must be with at least the *capacity* for culture, as the above authors for the most part acknowledge. Even if cultural change outraces organic evolution, creating blinding confusion through environmental novelties, to view the significance of its changes and its traits as independent of natural selection of genetic alternatives, or as mere thwarters of it, would be parallel to supposing that the function of an appetite is obesity (for a parallel argument, see Irons, 1979b).

It seems to me that the important question about cultural evolution is: Who or what determines which novelties will persist, and how is this determination made? On what basis are cultural changes spread or lost? In other words, we must analyze exactly the same part of the process of cultural change as for genetic change. For culture the answer to this question of who or what decides, and how it happens, actually determines the heritability of culture, since heritability of cultural items at least theoretically can vary from zero to 100 percent, whether across generations or even within a generation. Unlike a gene, a cultural trait can be suddenly abolished, and just as suddenly reinstated, across the whole population. At least in theory, this can be done by either conscious decision or some kind of unconscious process, as a result of what the involved parties see as their own best interests at the time. This reinforcing relationship among selection, heritability, and mutations in culture means that in cultural evolution, unlike organic evolution, heritability of traits will not be steadily increased and mutability depressed, because most mutations are deleterious to the individuals in whom they arise owing to the lack of feedback between mutational directions and adaptive value. Some cultural changes (mutations), at least, appear (i.e., are implemented, or translated from thought to action) because they are perceived ahead of time (while they are no more than ideas or perceptions in the minds of their creators) to have value; whether by plagiarism, teaching, or imitation, their spread is probably most often for this reason (e.g., see Griliches, 1957, and Hamblin and Miller, 1977, for analyses of the spread of adoption of hybrid corn by mid-western farmers in the United States). The reason for the failure of theories of culture based on rational strategizing are that (1) function was assessed at group levels, and it was clear that the best interests of the group were not always being served by any such rationality, and (2) the complex effects of individual and genic differences of interest were never adequately taken into account (Alexander, 1979a).

Unlike genetic evolution, then, cultural change involves a positive feedback between need and novelty. Whereas genetic evolution is characterized by a lowered rate of mutations, cultural inertia derives from the conflicts of interest among individuals and subgroups, from power distributions that result in stalemates, and from the incidental long-term persistence of some cultural institutions. This means that culture, as a whole, will not be seen as maximizing group interests. The reinforcement between need and novelty, together with efforts by individuals and subgroups to modify culture in their own interests, means that cultural change may be expected to continue accelerating, compared to rates of genetic evolution. This divergence, I believe, will make it increasingly difficult to interpret current human behavior in terms of history, because the acceleration of rates of appearance of novelties will decrease the abilities or tendencies of individuals to follow their own genetic interests. The acceleration of cultural change will also increasingly become apparent as the source of novel ethical problems, which are bound to increase in number and severity as cultural change accelerates. Ethical problems derive from conflicts of interest, and, as technology and social structure become more complex, conflicts of interest must also become more numerous and complex. In recent years novel ethical problems have begun to appear almost daily in the popular press; they involve topics like euthanasia, abortion, birth control, food additives, permissible levels of pollution, drug use, safety levels in automobiles, tolerance of divergent life styles, discrimination against minorities, and freedom from surveillance.

Most recent and current efforts to relate genetic change and cultural change, it seems to me, have actually been efforts to divorce them—to explain why and how culture and genes came "uncoupled" during human history. These arguments generally assume that such uncoupling is essentially synonymous with the appearance of culture—that culture is, by definition, a disengagement of human behavior from the effects of genes.

For those who regard the advent of culture as little more than an uncoupling of human behavior and its effects from the differential reproduction of genetic alternatives, there seems to be no reason whatever for a connection between the nature of cultural changes and the causes of selection of genetic replicators. They see what Cloak (1975) called "cultural instructions" and what Dawkins (1976) called "memes" or "cultural replicators" (personal communication) as changing independently of the effects of natural selection on genetic replicators. This view seems faulty to me. There is every reason to expect a correlation between cultural change and inclusive-fitness-maximizing; if none had existed the capacity for culture could not have evolved by natural selection of genetic alternatives (Alexander, 1971). To whatever extent the use of culture by individuals is learned—and if this is not the rule then one is at a loss to explain how any special human capacity to use and transmit culture could have evolved—*regularity of learning situations or envi-*

ronmental consistency is the link between genetic instructions and cultural instructions which makes the latter not a replicator at all but, in historical terms, a vehicle of the genetic replicators.

In whatever sense, and to whatever extent, learning is not a "blank slate" phenomenon in usual human environments—and surely it is not—then culture is a vehicle of the genes, and the startling matches of cultural patterns to predictions from inclusive-fitness-maximizing (Alexander, 1977b; papers in Chagnon and Irons, 1979) are in no way accidental or incidental effects of our genetic and evolutionary past.

I can think of no better illustration of the above point than the clear evidence that the basis for avoidance of sexual relations between very close relatives such as siblings (i.e., incest avoidance) is not their genetic relatedness as such but the fact that they are consistently reared together, thus socially intimate while prepubertal. Shapiro (1958), Talmon (1965), and Shepfer (1971, 1978) show that among a large number of Israeli children reared in situations quite similar to those of siblings *none has ever married within the kibbutz!* Similarly, Wolf (1966, 1968) presents evidence that among Taiwanese the practice of adopting the bride into the groom's family and rearing the two together as if they were siblings reduced the success of the marriage. This could only happen if incest avoidance were *learned,* and unless someone can bring forth an argument denying the universally accepted adaptive value of genetic outbreeding, this learning of incest avoidance occurs in a specific and narrow direction that is advantageous to genetic reproduction.

Richerson and Boyd (1978) argue, somewhat differently, that different optima exist for cultural and genetic reproduction, and therefore that their separate modes of inheritance lead to an uncoupling of their directions of change. They may base such arguments partly on examples such as teachers or priests who spend their lives imparting particular cultural traits to others at the apparent expense of curtailing or foregoing genetic reproduction. This they would apparently see as a situation in which, as Richerson and Boyd put it, the "culture-type" has become more important than the genotype. Two reservations about such interpretations seem necessary. First, teachers, or spreaders of cultural traits, may be using this activity as an indirect vehicle for the transmission of genes, just as anyone may so use any way of making a living or acquiring wealth, status, and power to assist his descendants and other relatives. Second, celibacy may occur as a result of environmental change causing a genetically nonreproductive *effect* from a previous genetically reproductive *function:* some modern celibates may be acting out a scenario that in an earlier or different time led to much reproduction, even via direct descendants, whether or not it now leads to any such consequence. Indeed, the solemnization of celibacy rites may sometimes be linked historically to public fears about the positions of power and opportunity afforded exalted religious leaders, from which unusually broad avenues of reproductive possibilities could derive. In some societies of the world religious leaders are in fact allowed unique and extensive sexual privileges (Murdock, 1949).

Cultural novelties do not replicate or spread themselves, even indirectly. They are replicated as a consequence of the behavior of the vehicles of gene replication. Only if the decisions or tendencies of such vehicles of gene replication (individuals) to use or not to use a cultural novelty are independent of the interests of the genetic replicators can it be said that cultural change is independent of the differential reproduction of genes. There is no reason to suspect such an independence, and every reason to expect the opposite, except, paradoxically, in a culture in which individuals are conscious of their own natural history.

The independent acceleration of the rate of cultural change in relation to that of genetic evolution, and the resultant possibility of altering the human social and developmental environment through consciousness of the mechanisms of its history, are sufficient to disconnect culture and genes. Without conscious knowledge of the reasons for our social history, however, even massive environmental novelty may not be enough to eliminate the connections between genetic and cultural change.

I think these are the reasons why efforts to understand culture in biological terms have so far failed. We can easily assume that the capacity for culture has allowed (as an incidental effect) various degrees of uncoupling of human behavior from patterns that would maximize genetic reproduction. In modern urban society, such uncoupling is rampant. It is my strong impression, for example, that the use of self-reflection to contemplate the *raison d'être* of genes is not an evolved function of these genes but an incidental effect of their action which owes its existence to the rise of technology. It is an effect, moreover, which I believe is less likely to serve than to thwart the differential reproduction of the genes we now possess. In other words I suspect that we would have to undergo considerable genetic evolution before reflection on our genetic history would be likely to cause us to maximize our reproduction.

There is enough evidence, even in everyday life, to indicate that human social behavior is usually patterned so as to correlate remarkably well with survival, well-being, and reproductive success. If one accepts this evidence (some of which is presented in the next chapter), then the real question becomes: What forces could cause the continued *coupling* between culture and genes? In effect, we must discover, for cultural evolution, the nature of the "hostile forces" (paralleling Darwin's "Hostile Forces of Nature", responsible for natural selection's effects on gene frequencies) which differentially favor variations in human social behavior and capacity through the adjustment by individuals and groups, consciously and otherwise, of strategies or styles of life.

Few people would doubt that positive and negative reinforcement, and responses to reward and punishment, are functions of environmental phenomena reinforcing (1) survival and well-being, and (2) avoidance of situations deleterious to survival and well-being. With ordinary physical and biotic stimuli this relationship is easy to understand: we withdraw from hot stoves, avoid poisonous snakes, seek out tasty foods, appreciate warmth in winter, dislike getting wet in cold rains, etc.

What about social stimuli? Should it not be the same? Should we not seek social situations that reward us and avoid those that punish us? Should not the actual definitions of reward and punishment in social behavior, as with responses to physical stimuli, identify for any organism those situations that, respectively, enhance and diminish its likelihood of social survival and well-being, and in turn its likelihood of reproductive success? Is it not possible that Sheldon (1961) was right in suggesting that "the reason why many pleasures are wicked is that they frustrate other pleasures"? That evil consists "in frustrations, as the Thomist says, in privation of one good by another"? Is what is pleasurable, hence seems "good" and "right," that which, at least in environments past, maximized genetic reproduction? I suggest so, even to the degree that our pleasure depends upon a capability of judging reactions of others to each of our actions, including the use of planning, self-awareness, and conscience to establish our own personal guidelines to success in sociality. I shall return to these topics later.

Arbitrariness in Culture

The symbolic or seemingly arbitrary nature of many cultural traits is commonly viewed as evidence against any functional theory, and especially against the notion that culture can somehow be explained by a history of differential reproduction by individuals. Of course, what seems to be arbitrariness may not be, and may only represent failure by the observer to understand the effects of different environments upon culture. Actual arbitrariness may be a relic due to the inertia of culture in the face of environmental shifts; or it may be a mistake about what kind of behavior will best serve one's interests—especially in the face of the accelerating introduction of novelty, primarily through technology. But, even if some trait is actually arbitrary, this need not be evidence against a theory based on inclusive-fitness-maximizing, particularly if culture is explained as a product of the different, as well as the shared, goals of the individuals and subgroups who have comprised human society during its history. Thus, however symbolism and language arose—say, because they were superior methods of communication—as major sources of arbitrariness they have also allowed the adjustment of messages away from reality in the interests of the transmitting individual or group. In other words, as abilities and tendencies to employ arbitrary or symbolic meanings increased the complexity and detail of messages, and the possibility of accurate transmission under difficult circumstances (e.g., more information per unit of time or information about objects or events removed in time or space), they also increased opportunities for deception and misinformation. As a result, some directions of cultural change may have been arbitrary in all regards except in their effects on the reproduction of their initiators and perpetuators.

Consider the relationship between status and the appreciation of fashion, art, literature, or music. What is important to the would-be critic or status-seeker is not alliance with a particular form but with whatever form will ultimately be regarded as most prestigious. If one is in a position to shape opinion he can, to one degree or another, cause it to become arbitrary. Fashion designers, the great artists, and the wealthy continually use their status to cause such adjustments. In no way, however, does such arbitrariness mean that the outcomes are trivial or unrelated to reproductive striving. Precisely the opposite is suggested—that arbitrariness may often be imposed, in regard to important circumstances, because the different outcomes are crucial, and because imposing arbitrariness is the only or best way for certain parties to prevail. *Alliance with a powerful or influential person's opinion is not arbitrary, even if the opinion itself is.* I believe that this point is crucial for the understanding of much about linguistic change, which is often touted as evidence of arbitrariness in culture. Sometimes a new word or combination may represent a refinement in communication that is of obvious value to those who hear and perpetuate it. At other times, though, whether or not such a change persists may depend largely or solely upon its source: there is little doubt that the use of the verb "finalize" by President Dwight D. Eisenhower accelerated its incorporation into everyday American usage.

Thus, when Hoebel (1954, p. 291) remarks that "no linguist could predict in advance what the language of an unknown people will be," he refers to an uncertainty based, not on some kind of absence of adaptive significance for linguistic change, but on an ignorance of the very special circumstances that surround adaptiveness in linguistic change. These circumstances involve the individual peculiarities of persons of differing power and influence, and the uniqueness of different social situations. There is no more reason to translate this kind of failure of predictability in linguistic change into an assumption of adaptive neutrality or special mystery than there is to make the same assumption for long-term or geologically ancient trends in genetic change, because the environmental conditions accounting for differences in adaptive significance there are also unknown.

These arguments may explain both the genesis of theories that great men are responsible for the patterning of culture, and the failure of such theories as general explanations. Great men do appear, and their striving, almost by definition, is likely now and then to have special influences; but, for reasons given above, these are not necessarily great influences nor influences leading to particular, predictable, overall changes in culture. Individuals may be perceived as acquiring greatness not only for lasting changes they bring about in respect to culture, but as well for any extraordinary ability to use or manipulate culture to their own ends, without causing lasting changes. That individuality of striving occurs within culture but does not necessarily lead to trends is also emphasized by the old saw in anthropology that "one hen-pecked husband in a village does not create a matriarchy." Similarly, the argument about status and arbitrariness is a variant of the adage that "when the king lisps everyone lisps," which merely satirizes the so-called "trickle-down" effect in stratified or hierarchical social systems. But it indicates that the "trickle-down" effect, rather than being a societal "mechanism for maintaining the motivation to strive for success, and hence for maintaining efficiency of performance in occupational roles in a system in which differential success is possible for only a few" (Fallers, 1973), is a manifestation of such striving, and a manifestation of degrees of success.

Arbitrariness, then, in fashion or any other aspect of culture, may not be contrary to the genetic reproductive success of those initiating and maintaining it, only to that of some of those upon whom it is imposed, in particular those who are least able to turn it to their own advantage. To understand the reproductive significance of arbitrariness as a part of status-seeking, one need only understand the reproductive significance of status itself. One might suggest that there are genetic instructions that somehow result in our engaging in arbitrariness in symbolic behavior in whatever environments it is genetically reproductive to do so.

In summary, I suggest that the rates and directions of mutability and heritability in culture are determined by the collective interests and compromises of interest of the individuals striving at any particular time or place, together with the form and degree of inertia in the cultural environment as a result of its history; that the "hostile forces" resulting in cultural change have tended increasingly to be conflicts of interest among human individuals and subgroups in securing relief from Darwin's "Hostile Forces of Nature"; and that, among these "Hostile Forces of Nature," increasingly prominent and eventually paramount have been what amounted to predators, in the form of other humans acting as individuals, or in groups with at least temporary commonality of interests.

These arguments lead to four expectations: (1) a reasonably close correspondence between the structure of culture and its usefulness to individuals in inclusive-fitness-maximizing, (2) an even closer correlation between the overall structure of culture and those traits that benefit everyone about equally, or benefit the great majority, (3) extremely effective capabilities of individuals to mold themselves to fit their cultural milieu, and (4) tendencies for culture to be so structured as to resist substantial alteration by individuals and subgroups in their own interests and contrary to those of others. These predictions lead us to analyze the variations in culture as the potential outcomes of different strategies of inclusive-fitness-maximizing behavior in different circumstances. They also suggest that the proximate (or immediate) physiological and social mechanisms whereby inclusive fitness is maximized are potential explanations of the degrees and directions by which cultural patterns diverge from actual inclusive-fitness-maximizing behaviors when technological and other changes rapidly create evolutionarily novel environments.

The Problem of Individual Development

It is one thing to suggest that humans are like other organisms in that individuals evolve to behave so as to maximize their reproduction; it is quite another to understand exactly how such effects could be achieved in ways that do not violate our understanding of the nature and importance of learning in the development of social behavior.

It is often argued, or assumed, that the human style of social learning and the traditional transmission of culture cause human social behaviors to be too remotely and too indirectly tied to gene action, and thus to inheritance through genes, for evolutionary theory to predict anything significant about the structure of human sociality or culture, or the directions of its change. From this stance it might even be argued that apparent correlations between Darwinian predictions and social or cultural patterns are necessarily fortuitous.

It is also widely believed that biological theories of human social behavior demand a kind of ontogenetic determinism unacceptable to most and repugnant to practically everyone—that an evolutionary view of human behavior necessarily violates concepts of social learning, free will, justice, and common sense. But this need not be so. I see no reason why the physiological or proximate mechanisms of inclusive-fitness-maximizing social behavior in humans should not be entirely compatible with the idea of free will. Indeed, I will argue instead that a study of these mechanisms is the only reasonable means for understanding phenomena like free will, justice, happiness, consciousness, learning, purpose, and the distribution and nature of altruism (both consciously intended and nondeliberate), and all of the other special human behavioral phenomena.

All traits are products of the interaction of the genetic materials with the environment; one cannot produce an organism of any kind with either alone. Darwinism, however, says nothing about the directness, indirectness, or complexity of the routes that must exist between gene action and behavior. It does suggest that the ontogenies of organisms should evolve such that organisms tend to learn, or to develop abilities, to reproduce maximally in their particular environments, and that they should not develop any other abilities. And it implies that the routes existing between gene action and behavior, whatever they are, will be those most appropriate to the patterns of predictable change characteristic of the succession of environments in which the trait has been expressed throughout history. If Darwinism is the correct theory of evolutionary history, it must account for all learning, and although it has not been widely considered in this context, it should lead to more useful predictions about learning than any other kind of theory.

Let us, then, consider in some detail the general question of ontogenies or developmental pathways, and how they evolve, before taking up the particular question of proximate mechanisms of inclusive-fitness-maximizing through nepotism.

EVOLUTION AND LEARNING

Ontogenies are the most difficult and perplexing phenomena that evolutionary theory must account for, and behavior is the most perplexing of all phenotypic attributes, in ontogenetic terms, for it is less directly related to gene action than its underlying morphological and physiological correlates (Alexander, 1971, 1975a). Especially difficult to understand is the evolutionary background of learning, as is illustrated by the fact that nearly all theories of learning are concerned with physiological and ontogenetic mechanisms and sequences, scarcely any with adaptive (reproductive) significance. As such, learning theories carry no probability of accounting for the presence and nature of ontogenetic or physiological mechanisms, or even for different mechanisms in different kinds of organisms; hence, perhaps, the tendency of learning theory to revert to singularity and, paradoxically, to seem more general than it is.

The essence of learning is adaptation to immediate contingencies. The widespread existence and elaborateness of learning, therefore, testifies to the absence of long-term causation in the environments of life, and therefore the value of reliance upon immediate contingencies. A paradox, in exploring learning as an evolutionary adaptation, is that the more prominently an organism's phenotype becomes responsive to immediate contingencies the more difficult it is to explain the phenotype by reference to genetic or evolutionary history. It is easy to forget that whenever responses to immediate contingencies fail to maximize reproduction the flexibility which permits them will be counterselected. Phenotypic flexibility does not mean that there is no genetic command as to which responses will result from which environmental contingencies. Nor does an absence of heritability in phenotypic variance because selection has been inexorably directional imply that there is no genetic contribution to the expression of the trait. The commands given by genes are for the production of given phenotypic responses in given environments. So far, with humans we have nearly always countermanded by altering the environment in which the trait is expressed, ranging from drug, hormonal, and other chemical therapy to deliberate modifications of teaching programs and social experience. Thus, even the clearly genetically determined phenotypic differences between males and females (except for the production of eggs and sperm) can essentially all be not merely erased, but *reversed,* by changing the individual's hormonal and social environment (Money and Ehrhardt, 1972). It is likely that we will some day be able to alter genic commands as well, not merely by selective breeding (practiced to some degree now by genetic counseling before marriages or offspring production for the benefit of people with potential problems known to derive from genetic variations such as with sickle cell anemia and cleft palate), but by actually altering either the genes or their environment of expression.

To some extent the ontogenetic or experiential backgrounds

of ordinary human behavior may actually seem less inscrutable than the backgrounds of other traits, or of much of the behavior of nonhuman species. The essential stimuli for altering human behavior, after all, are often transparent. On this account, an evolutionary explanation may seem superfluous. Nevertheless, because particular stimulus sequences lead predictably to particular behavioral responses, all possible responses are evidently not equally probable. Before evolutionary approaches can be regarded as useless, one must know the relationships among ontogenetic stimuli, behavioral responses, and the routes to reproductive maximization in the relevant environment.

The Heritability Problem

It is sometimes argued that unless one can show that the behavioral (or other) traits he studies vary, and the variations can be shown to be heritable (that is, to be associated with genetic variations), it is futile to argue that natural selection underlies the expression of the traits. Controversies over this issue have been almost unbelievably complex and protracted (e.g., Feldman and Lewontin, 1975). When the question involved is whether or not variations in some particular human trait are heritable, such as differences in scores on intelligence tests (which have been used to help decide the fates of individual humans), the reasons for demanding extraordinarily convincing evidence are obvious; indeed, the potential for misuse and self-serving interpretations by those in positions of power may sometimes seem to justify the rejection of any evidence whatever. But the arguments have been extended beyond such cases, almost to deny evolution itself. It is important here to take a brief look at the most general aspects of this controversy.

Evolution by natural selection can occur only when the phenotypes of the individuals in a population vary, and those variations correlate not only with genetic differences, but also with variations in reproductive success. The potentially confusing consequence that is sometimes exploited to denigrate the significance of natural selection is the following: if selection is consistent in a given direction it tends to eliminate genetic variations that correlate with variations in reproductive success. This leads to a situation that seems paradoxical for two reasons. First, any remaining phenotypic variation that relates to reproductive success in the context being examined is unlikely to correlate with genetic variations. Second, any remaining genetic variations will not correlate with differences in reproductive success in the context being examined. The reappearance of evident genetic variations relevant to reproductive success in the context being examined will depend on either beneficial mutations, which are exceedingly rare (deleterious mutations are less likely to be evident because they will not spread), or a change in the direction of selection, which may also be rare. Either of these effects can also be fleeting and difficult to detect if the resulting selection is quite effective.

And so, in biology, we are continually faced with the paradox of living organisms, supposed to have evolved by natural selection, yet, because we observe them only across brief time spans, seeming to lack the very attributes described by evolutionists as prerequisite for the process to occur. The confusion is furthered by two other facts. First, it is very difficult to measure reproductive success in an evolutionarily meaningful fashion. Should one count offspring, and if so at what stage? Or should one count grandchildren, or try to measure effects on all kinds of genetic relatives? The second problem is that all organisms are to one degree or another phenotypically plastic or variable, in ways that enhance reproduction in the different environments to which they are commonly exposed; such plasticity is almost the definition of phenotype. The potential for each of these phenotypic variations to be expressed may depend upon a specific genetic background as much as the different variations depend upon specific sets of environmental stimuli. Nevertheless, even highly evolved variations, uncorrelated with genetic variations (such as behavioral differences deriving from the universal possession of complex learning ability), tend to further the illusion that organisms do not possess the features necessary for evolution by natural selection.

All of these confusing features of organisms are logical, predictable, even inevitable outcomes of evolution by natural selection. Yet by their nature they also make it easy to deny the significance of natural selection, particularly to an audience that already wishes to disbelieve on other grounds, such as religion or politics. This is especially true for the phenotypically most malleable organisms, such as humans, and for the most malleable aspects of the phenotype, such as behavior.

These same features of life are the source of most of the confusion about heritability. As a practical fact, heritability is measurable only by examining the correlation of genetic variations with phenotypic variations. But natural selection removes such variations when they correlate with variations in reproductive success. As a consequence, and paradoxically, traits may become increasingly heritable (i.e., less influenced by environmental variations) as the only practical measures of heritability decrease and disappear.

The above features of evolution by natural selection are not theoretical. They are facts, demonstrated countless times in the selective breeding of animals and plants, and directly observable in nature. As a result it is obviously unreasonable to allow the problem of measuring heritability in any given case (or the use or misuse of supposed heritability or its absence in the variations of some socially significant human trait) to cast doubt upon the centrality of natural selection of genetic alternatives in producing life and its traits.

The real question is: what actually is inherited? We can understand the significance of the problem by realizing that even if directional selection tends to remove heritable variations it does not follow that universality or nonvariability in traits necessarily means that such traits have somehow lost their ontogenies—or potential for varying—or become, in some fashion, "innate," "inborn," or "instinctive." Avoidance of mating with close relatives is an excellent example. As noted elsewhere in this book, in many organisms outbreeding is easily shown to be the result of social experience. Mice and men alike tend to avoid sexually not their close genetic relatives but those with whom they have been socially reared, whether relatives or not, and they mate more successfully with those with whom they have not been reared, again whether relatives or not (Hill, 1974; Wolf, 1966, 1968; Shapiro, 1958; Shepfer, 1971).

These results, on the other hand, do not mean that nothing is inherited that is relevant to either genetic outbreeding or sexual avoidance of those with whom one is reared. On the contrary they indicate that something that is inherited is relevant to both outcomes. That something is certainly genes that, in some fashion (still unknown despite decades of extensive and intensive analyses of learning by a science of psychology impressed with the centrality of incest avoidance—e.g., Lindzey, 1967), give rise to ontogenies culminating in sexual avoidance of close social associates during juvenile life. Thereby, in the usual human environments, these same genes incidentally bring about genetic outbreeding. Nevertheless, it is almost certainly the positive reproductive effects of genetic outbreeding that caused the spread of these genes whose action we still do not understand, and whose presence we can yet verify only indirectly.

Heritability, then, is a feature of life that is basic to evolution. What is actually inherited during evolution is, first of all, genes, and whether or not the traits to which genes give rise during ontogeny are consistent depends upon the consistency of the developmental environment. What constitutes consistency in the environment (from the organism's point of view) in turn depends on the history of environments, and the effects of that history, generation by generation, on which genes are saved and collected together.

To summarize, we know very little about the ontogenetic basis of heritability in phenotypic traits. But we do know, in general, how ontogenies and heritability both fit into the evolutionary process, that ontogenies do not disappear as the heritability of traits increases (even though they may shorten and simplify), and therefore that traits that are highly uniform in one environment (hence, may be regarded as highly heritable) may also be highly changeable in another.

Superseding Biological Constraints on Human Behavior

Appropriate ontogenies, in selective terms, are those leading to maximal genetic reproduction. This does not imply, however, that every act of every human is to be interpreted as maximizing its reproduction, but that evolutionary change, in the history of environments in which humans have lived, has tended to be in the direction of maximizing reproduction. The environment in which human behavioral traits develop includes not merely external physical and biotic stimuli but internal changes like knowledge of human history, tendencies, and motivations, and the probable rewards and punishments for given acts. Whatever the extent or nature of biologically based constraints on the modifiability of human behavior, therefore, such constraints seem most likely to be effectively bypassed or superseded by humans who individually and collectively are aware of them and understand them well. Behavioral constraints and plasticities are each to be understood only in terms of the particular environments in which they apply, and these environments are probably best identified from a thorough understanding of both proximate and ultimate backgrounds of behavior. Indeed, the most significant change possible in one's environment of behavioral development may be the introduction of new ideas and knowledge which improve one's ability to reflect on one's own tendencies and motivations, and those of one's fellows.

For example, should I suggest to an acquaintance that he is compelled to do in any situation what will be for him most reproductive, he might scoff and invite me to identify such courses of action so that he could prove me wrong by repeatedly doing the opposite. If the opposite act were very expensive or detrimental, he might prefer to bet me that he would do the unexpected. Upon doing so he would triumphantly take ownership of whatever I had lost in the bet. He might not even consider the significance of the use to which he would eventually put his winnings, or exactly how he arrived at the size of the incentive necessary to make him take the particular course of action required to win the bet. Moreover, it would be easy for him to neglect considering what it means that the intensity of our argument, or the number and significance of the persons observing our disagreement, might also influence his decision about which course of action to take.

Human reproduction is realized through devious, complex routes, and its currencies are numerous, variable, and sometimes unrecognized as such. The important point is that my statements to the person in this example, as well as all of the circumstances created by our actual disagreement, would be a part of the environment in which he made his cost-benefit decision about which course of action to take. A history of differential genetic reproduction is thus most deterministic for the human still unaware of it.

Needless Sources of Controversy

The recent rise of attention to the new view of natural selection has shown that the relationship between natural selection and phenotypic plasticities, as expressed by learning and culture, is one of the most poorly understood of all biological phenomena. Moreover, this relationship seems to be especially poorly understood among social scientists and humanists, the very area in which it must be clarified if evolution and human nature are to be correlated accurately. The ignorance exists partly because those who study the ontogenetic basis of human behavior have only rarely studied evolutionary mechanisms, and also because evolutionists commonly leave the question of proximate mechanisms aside as long as they can. The consequence of this breach has been a great deal of emotionally charged and needless controversy, and sometimes an almost complete lack of understanding between biologists and those concerned directly with human behavior.

An excellent example occurred in an exchange between the

biologist Edward O. Wilson and the anthropologist Marvin Harris (Harris and Wilson, 1978). Harris had just asserted that investigators such as Trivers, Dickemann, Chagnon, and I were "concerned with explaining the social response repertories of different cultures in terms of differences in gene frequencies." No words have been written, nor I suspect spoken, by any of those four people which would substantiate Harris' statement in even the slightest degree, and I cite my own papers (especially Alexander, 1977a, b, 1978b, 1979b) as explicit denials, together with the reasons for doubting what Harris says we believe.

A little later in the *Sciences* debate (p. 12) Harris contrasted cultural practices of the Nayar of South India and the Bathonga of Mozambique and asked Wilson if the two patterns could derive from populations with the same gene frequencies. When Wilson replied "That is a good possibility," Harris responded, "Then in what sense is that biology?" The error is clear. Harris has to suppose that the ability to behave differently in different circumstances with the same set of genes cannot be a product of natural selection. This belief, or attitude, is very widespread. It is, unfortunately, a consequence of a profound misunderstanding of biology and evolution. If there is one thing that natural selection has given to every species it is the ability to adjust in different fashions to different developmental environments. This is what phenotypes are all about, and all organisms have phenotypes. If there is an organism most elaborately endowed with flexibility in the face of environmental variation, it is the human organism. Thus, it *is* biology for the Nayar and the Bathonga, and all the people who live elsewhere in other kinds of cultural and ecological environments, to adjust and develop and learn their behavior so that it is expressed appropriately for each of the special circumstances in which they live and reproduce. What we do not yet know, in each case, is the sequence of prior social, ecological, and other circumstances that have set all of the different patterns.

Several sources of confusion about evolutionary biologists' views of behavioral ontogenies and the notions of "instinctive," "innate," or "genetically determined" behavior derive from their approach to the study of traits. In general, evolutionary biologists proceed as follows: first they identify phenotypic traits in organisms, then they study the adaptive (reproductive) significance of those traits, almost as if the traits had no ontogenies—as if there were no proximate physiological or developmental mechanisms on which they depend. Evolutionists postpone the analysis of ontogenies because their initial interest is only in how the trait is expressed in the usual environment of the species, which is what determines its evolutionary adaptiveness. Moreover, evolutionary biologists also tend to focus more on whether or not variations in traits correlate with genetic variations than on whether or not they correlate with environmental variations, because they are interested in whether or not natural selection can operate with respect to the variations.

For evolutionary biologists, genes have been studied largely as recombining units of heredity; hence their effects have been described chiefly as *differences,* and always as *phenotypic* differences. This approach has also caused evolutionists to ignore ontogenies and environmental determinants of phenotypic differences. One justification for the evolutionist's approach is that it corresponds to the way in which natural selection works. It does not matter, in selection, how a trait comes to be expressed—only that it is expressed in the optimal way for genetic reproduction, and at the optimal time and place. The reasons ontogenies vary for different organisms and different traits are simply that, first, different kinds of ontogenies deliver particular phenotypic responses more or less reliably, and, second, natural selection always begins with last year's model, and the previous lines of specialization may have caused different routes of further specialization to be more or less likely.

It is probably worthwhile to point out that cultural anthropologists follow almost exactly the same procedure as that described above for evolutionary biologists: they first describe, tabulate, and compare cultural variations and only subsequently seek out their immediate bases. The proximate (physiological, morphological, behavioral) background for any behavior is extraordinarily difficult to explicate; probably a complete sequence has not yet been worked out for a single behavior of any organism. The biologist assumes that this deficit is owing to ignorance, and that connections between natural selection and ontogenetic patterns really do exist whether or not we know what they are. Some anthropologists, however, for perhaps understandable reasons, have found it easy to doubt that connections exist at all between culture and genes—between cultural variations and natural selection. At the very least I think they should want to know what is predicted by the new approach in evolution before they continue to behave as though organic evolution could not have shaped the capacity for culture and thus the actual expression of culture in different circumstances.

The evolutionary approach of deferring the scrutiny of ontogenies is perfectly all right for nonhuman organisms. Suppose that one makes a few predictions from an inclusive-fitness model about how a group of frogs will behave in a pond, and he tests the predictions and discovers that the frogs' behavior does indeed meet them. The people who learn of this analysis may accept the arguments, reject them, or draw no conclusions. All of those people, however, will probably work up some kind of mental image of what the arguments about the frogs' behavior mean with regard to the ontogenetic background of the behavior. The chances are very good that, whether they accept the arguments or not, their views of how the frogs' genes influence their behavior will be oversimplified.

After all, the idea that behavior is determined by the genes is simpler than the idea that behavior is determined by the environment, or by the genes and the environment. To say that behavior is determined by the environment, or by the genes and the environment, does not say much of anything, because the next question is: how? To say that behavior is determined by the genes seems to settle everything.

One might say: So what if everyone oversimplifies the ontogenetic basis of frog behavior? It matters, but it is not crucial.

After all, no one thinks up social and educational programs for frogs. But they do for humans. And what they think up depends upon what they see as the respective roles of genes and environments in the ontogeny of human behavior.

This realization came only in my reflections on what I saw as repeated misinterpretations of discussions of human behavior in an evolutionary context. I had always regarded Darwinism as essentially a way of interpreting history, rather than as a basis for ideology. But if its application to human affairs is widely misinterpreted or misused, then it unavoidably becomes ideological in its effect even if not in the purpose of its practitioners. If one sees one's self as an anti-determinist, then it is very easy to be an unreasonable anti-evolutionist; if one sees one's self as an evolutionist then it is very easy to be an unreasonable determinist.

Now we can understand, finally, why people like Marvin Harris and Sherwood Washburn (1978) believe that biologists who speak of human culture are "genetic determinists" and why they automatically suppose that we believe that human cultural patterns vary because the genetic composition of human populations varies. It is not because they have discovered that we all *say* such things, but because they actually equate "biological" with "genetic" and believe that if we speak of a biological function or background for cultural variations we can only mean a background in genetic variations. Unfortunately this particular kind of misunderstanding, or its use, is not peculiar to nonbiologists. I admit to perplexity when I see the same arguments being used by biologists such as Slobodkin (1977) and Boucher et al. (1978).

THE VARIOUS MEANINGS OF DETERMINISM

Most of the people with whom I have discussed determinism accept that, for most purposes, events can be viewed as having been preceded by continuous chains of causation. Whether or not, in this sense, events are "determined" is not the central issue in explaining the traits of living organisms, although it

may well be an important issue in s[
physicists and philosophers about the l[
particles. Sometimes it seems that th[
things which involves determinism is,[
of presumed causal events variously[
trait under consideration. Thus, *genetic* determinism ....p—
the genes received by an organism can absolutely determine
some aspect of its behavior, no matter what subsequently happens to the organism. The effect of this argument is to exclude
environment whenever environment is used, as I believe it is
generally used in biology, to mean all contingencies other than
genes; so it is a ridiculous argument. This is just as true for the
case of "pathologies" induced by single gene mutations (e.g., E.
O. Wilson, 1978) as for any others; even these genes realize
their effects only in certain environments, and the possibility
always exists that we can create an environment in which the
effects do not occur.

Evolutionary determinism, on the other hand, is in a sense even
more remote in time than genetic determinism, because it includes events of natural selection that established or fixed certain genes long before they were received by the members of
any particular generation. But evolutionary determinism necessarily includes both genes and environment since natural selection is the effect of the environment on gene frequencies; and
this means that the "evolutionarily determined" traits of any
organism are actually alterable by modifying the ontogenetic
environments of individuals. Sometimes, genetic determinism
is used when evolutionary determinism is meant; but the two
are quite different. It is evolutionary not genetic determinism
that enables us to predict sex ratios, relationships between sexual dimorphism and breeding systems, the role of learning in
uncertain environments, etc. And it is evolutionary not genetic
determinism that most evolutionary biologists think may be
useful in understanding and changing human behavior.

Biological determinism, also a common term in the recent controversies, is even more difficult to define. I suspect, however,
that its meaning is frequently interpreted by thinking of the

ctive "biological" as being opposed to "cultural" or learned," hence as "genetic" or "physiological" (the latter in turn seeming often to be translated as "genetic"). This phrase is especially unfortunate because it is vague and pejorative, yet seems to refer to biology in general. The implication is that biologists are by definition deterministic in some unsupportable fashion, and that what is usually involved is a narrow view of the importance of genes in guiding ontogenies. For the most part, biologists deal with nonhuman organisms and admittedly are likely to have inadequate views of the sources of culture and the ontogenetic flexibility of human behavior. Labels like "biological determinist," however, do not seem designed to remedy this problem but to create a climate of interdisciplinary hostility and distrust.

If genetic determinism is such an unsupportable concept, why is it still used—for example, in the often-repeated statement imputed to biologist E. O. Wilson (e.g., N. Wade, 1976) that human behavior may be "10–15 percent genetically determined"? Sometimes this happens because the arguments made here have not been carefully considered, sometimes because the writer means that variations in behavior relate to variations in gene frequencies. So one could very well speculate, for example, that 10–15 percent of the variations in human behavior relate to variations in genetic constitution. This may or may not be reasonable, but at least it is not inadmissible as an hypothesis.

In times when remarks about phenotypes were not taken so emotionally, and when thoughts about heredity were not so closely tied to things human and behavioral, geneticists fell into using a shorthand that translated "the proportion of phenotypic variations that results from genetic variations" to "the proportion of the phenotype that is genetic." I think it is reasonable to argue now that such shorthand, especially in regard to human behavior, is almost certain to be misinterpreted, hence is inexcusable.

Another source of confusion about determinism arises from the statement that some behavior "has a genetic basis." In one sense all behavior "has a genetic basis," that sense being that it also has an environmental basis. Since the truth of this dual causation of all behavior is so obvious, to suggest that some particular behavior "has a genetic basis" may be interpreted as meaning that variations in that behavior have a genetic basis. Such statements are not to be lightly applied to human behavior.

Some individuals seem to be telling us that the proper way out of the dilemma is to stop analyzing human behavior in evolutionary terms until we have solved the ontogenetic problems. I emphatically disagree, for at least two reasons: first, everything we know about evolution we have learned without fully understanding the ontogenetic basis of any behavior in any organism; and, second, the ways to analyze ontogenies fruitfully all too often become clear only as a result of evolutionary approaches. Ontogenies, after all, are also products of natural selection.

The alternative solution is to devise hypotheses about ontogenetic mechanisms which, even if incomplete and imperfect, will be testable starting points, and will demonstrate, if possible, that to apply Darwinian models to human behavior does not automatically require an intolerable determinism.

Our general ignorance of ontogenies, despite concentration on learning and other developmental theories in the social sciences and among a significant proportion of zoological behaviorists, indicates that arguments from ontogeny about the probable failure of Darwinian theory are likely to be vulnerable. Preliminary findings already show that patterns of culture match predictions from the modern version of Darwinian theory to a much more significant degree than they were thought to in the past (see Alexander, 1974, 1977b; Chagnon and Irons, 1979; Alexander and Tinkle, in press; and the next chapter), indicating that objections from ontogenetic arguments must be reexamined. Moreover, ontogenies in general, of whatever kind of trait, are largely unknown; yet scarcely anyone invokes the complexities or supposed indirectness of ontogenies to support a general rejection of an evolutionary background for morphological and physiological traits.

I share the concern of those investigators of human behavior who wish to excise all implications of genetic determinism and unsubstantiated claims of genetically based variations from considerations of human behavior. But I also reject any suggestion that either of these implications or claims are necessary concomitants of an evolutionary view of human behavior.

CULTURAL DETERMINISM

One final source of confusion about cultural change deserves mention, and it returns us to the discussion that began this chapter. White (1949) remarks that culture derives from culture —that whatever changes occur depend upon the preexisting structure of culture. His statement is a classic comment on what has come to be called "cultural determinism" (and on the independence of culture):

> Apart from theories of environmental determinism which considered merely the relationship between habitat and culture, all types of interpretation prior to the emergence of anthropology as a science thought of man and culture together; no one considered culture apart from its human carriers. With the advance of science, however, came a recognition of culture as a distinct class of events, as a distinct order of phenomena. It was seen that culture is not merely a reflex response to habitat, nor a simple and direct manifestation of "human nature." It came to be realized that culture is a continuum, a stream of events, that flows freely down through time from one generation to another and laterally from one race or habitat to another. One came eventually to understand that the determinants of culture lie within the stream of culture itself; that a language, custom, belief, tool or ceremony, is the product of antecedent and concomitant cultural elements and processes. In short, it was discovered that culture may be considered, from the standpoint of scientific analysis and interpretation, as a thing *sui generis*, as a class of events and processes that behaves in terms of its own principles and laws and which consequently can be explained only in terms of its own elements and processes. Culture may thus be considered as a self-contained, self-determined process; one that can be explained only in terms of itself. [P. xviii]

But the problem of which White speaks is in no way restricted to culture; rather, it is closely paralleled in evolutionary biology. In organic as well as cultural evolution the forces of change—the actual process of evolution—can only operate on the phenotypes of the current generation—on the existing forms. This is true, no matter how or in what directions these forces of change may themselves differ from one generation to the next. Sometimes it will be exceedingly difficult to understand past forces of change, or their products, and the further one explores into the past the more scanty will be the evidence. Yet the traits that exist in any one time or place in organic phenotypes or human culture place great strictures on the directions and rates of subsequent changes. In this sense both organic and cultural evolution can only be understood in terms of themselves. It does not follow, however, that the actual forces of change in either case—the processes of cultural and organic microevolution—cannot be well understood from the analysis of change during even a single generation. That we must forever explore the long-term histories of our cultural and organic pasts with imperfect data does not deny the ability to understand their processes thoroughly, and therefore to understand a great deal about how they must have operated to produce the forms available for our direct observation and analysis, and how they are likely to operate in the future. Thus, despite the distinctiveness of cultural inheritance, and the obvious linking of its expressions to group activities, one can very well be a "cultural determinist" and still accept that connections between the differential reproduction of genes and the patterns of culture have tended continually to be in states of maintenance or restoration throughout the whole of human history.

The Proximate Mechanisms of Inclusive-Fitness-Maximizing Behavior

In order to evaluate the probable proximate mechanisms, or ontogenetic and physiological backgrounds, of inclusive-fitness-maximizing behavior, we can divide it into two major categories: (1) assistance to one's own phenotype and (2) assistance to the phenotypes of genetic relatives. Our aim, of course, is to see if there are reasonable ways to interpret all human behavior as either actually maximizing inclusive fitness or else representing the surrogates of such behavior under dramatically or rapidly changed environmental conditions.

The two categories given above correspond exactly to the evolutionary biologists' division of lifetimes into somatic effort (use of resources and taking of risks for growth and development and the accumulation of energy) and reproductive effort (use of resources and taking of risks in acts that if unthwarted represent actual reproduction); the two categories may otherwise be described as the garnering of resources and the redistribution of resources (figure 5, and Alexander and Borgia, 1979). Resources, in turn, may be defined as the means whereby Darwin's Hostile Forces of Nature are resisted and combated. In other words, resources are the means whereby reproduction is effected and maximized; they are what an organism uses to counteract all of the environmental factors that threaten and limit reproduction: predators, parasites, diseases, food shortages, climate, weather, and mate "shortages."

Assistance to one's own phenotype at first seems simple and straightforward: we should do what is best to promote our own survival and nothing else. Of course it is not so, and for an evolutionary biologist the first clue might have been (but probably never has been) the obvious fact that we, as individuals, have not evolved to survive indefinitely; only genes (or polygenes, supergenes, chromosomes or other genetic replicators) seem to possess this attribute. Lifetimes, and the activities which govern the effects of environmental insults on their lengths, are instead programmed according to a variety of highly predictive patterns indicating that reproduction not survival is their *raison d'être*. Multiplication appears to be the vehicle of immortality for genetic units, and phenotypes are the mortal vehicles of this genetic multiplication. Indeed, individuals are the shortest-lived of all the units in the hierarchical organization of life (figure 3), and the lifetimes of individuals in most species are very short.

Since genetic reproduction in ordinary sexual organisms is accomplished solely through assistance to others we must conclude that all egotistic, hedonistic, self-serving, self-loving behavior is to be understood only through its likelihood of contributing—at least in the environments of the past—to someone else's welfare (i.e., to reproduction via others). This means that we must examine every detail of apparent egotism in terms of opportunities for either (1) use of its effects on the phenotypes of the egotists themselves, or (2) willingness of surrender to nonegotistical alternatives in the interests of genetic reproduction. Just as it is no accident that we withdraw from painful stimuli and approach pleasurable ones (or, respectively, reduce and raise their likelihood of recurrence), it is also no accident that we are pleased when our children and other relatives succeed, and dismayed when they fail, or that we are willing to sacrifice our own phenotypes to ensure their success.

Paradoxically, in humans parental and other nepotistic activities may often be easier to understand in terms of ontogeny and physiology than the various kinds of egotistical activities (figure 5). The reason is that we grow up imbedded in extraordinarily complex networks of kin (figure 4), and we possess uniquely keen capabilities of knowing their degrees of relatedness and their needs. As a consequence, the ways of learning to be altruistic to kin are so many and varied that the identification of truly egotistical acts (somatic effort), as opposed to nepotism, may be difficult. Even young children, long before the age of offspring production, have innumerable opportunities to be self-sacrificing in the interests of their close kin.

THE ONTOGENY OF NEPOTISM

The immediate backgrounds of nepotism can be approached in two ways. One could either analyze the actual interactions of relatives and then postulate appropriate mechanisms, or one could consider first the theoretically possible and probable mechanisms for identifying and helping relatives. I shall combine these two approaches.

First, two major classes of nepotism can be distinguished, which I shall call *discriminative* and *nondiscriminative* nepotism. The underlying mechanisms for these two postulated classes are likely to be different. Nondiscriminative nepotism can be explained by considering two possible reactions to other individuals encountered: tolerate or help all of them indiscriminately or fail to tolerate or help them, also indiscriminately. The first response is likely to be adaptive only when the likelihood is negligible that nonrelatives (or relatives more distant than usual) can interlope and gain by accepting nondiscriminative altruism yet giving neither altruism nor genetic representation in return. This situation can be illustrated by comparing the parental behavior of solitary forms such as the American possum with that of group-living species such as herding ungulates. Female possums with young accept foreign young readily (B. S. Low, personal communication) while herding ungulates such as sheep and cattle, as well as other group-living forms such as Adelie Penguins (Sladen, 1955) and Bank Swallows (Hoogland and Sherman, 1976), feed and protect only their own offspring and can be induced to adopt others only with considerable difficulty. The adaptive background for the difference is clear. Under normal conditions, possums suffer practically no risk of accepting unrelated young while group-living animals constantly suffer this risk (Hoogland and Sherman, 1976, suggest that the ability of Bank Swallows to recognize their own offspring occurs at just the time when errors become possible, when the young first begin to move between nest burrows; similar suggestions have been made for other species —see Hamilton, 1964; Birkhead, 1978). This difference between solitary and grouping forms, incidentally, is almost the precise opposite of what would be predicted from a group-selectionist, group-altruistic hypothesis, under which group-living would be seen as having yielded many opportunities during evolution for learning how better to help the group by contributing to the welfare of orphans and other unrelated

needy young. Under an individual inclusive-fitness-maximizing hypothesis we should expect adoptions only in groups of
very close relatives (e.g., lions—Schaller, 1972) or as parts of
systems of reciprocal altruism (see p. 48).

In other words, we should predict that indiscriminate nepotism is likely only when all associates are relatives of a single class, and when nepotism has some strong possibility of
increasing the reproductive success of the helped individuals.
It may also evolve when different relatives are encountered
but no means of distinguishing them can evolve, as I shall
argue is the case among full siblings, which only average
one-half alike in genes identical by descent (that is, any two
full siblings are not necessarily half alike). Aside from isolated parents and their offspring, indiscriminate nepotism
may evolve in clones in asexual species and in isolated
broods of siblings in sexual species, such as in caterpillars
that move in groups.

In contrast, indiscriminate (or complete) failure to tolerate or
help other individuals would seem most likely in (1) sexual
forms (because individuals will tend to be genetically distinct),
(2) forms that disperse readily (separating genetic relatives),
and (3) forms in which there are few opportunities anyway to
contribute to the success of others (as when parental care is
ineffective).

Among higher animals at least, discriminative nepotism is
probably much more widespread than nondiscriminative nepotism, partly because most social organisms interact with more
than one kind of relative, and partly because nondiscriminative
nepotism requires circumstances in which altruists must not be
jeopardized by interloping nonrelatives. Among all organisms
discriminative nepotism is probably most complex in human
societies, in which each individual is in social contact with a
wide variety of relatives of different degree and need, and in
which nepotism can involve many alternative courses of action,
with correspondingly varied possibilities for genetic returns.

Having distinguished two major classes of nepotism for
which proximate mechanisms must exist, we can now begin the

search for such mechanisms by considering first the more easily reconstructable aspects of the bases for our own discriminative nepotism. To do this, at least in part, we need only recall the patterns of our own experiences with different relatives. Much of the background of our individual and personal knowledge of relatives is known to us, and obviously learned from parents and other relatives, and from various kinds of associations. To the extent that we can discover how we learn to distinguish relatives and their needs and characteristics, we will have an idea of the actual proximate mechanisms of human nepotism. In other words, whatever enables us to construct a diagram of the sort in figure 4, whatever makes the labels on that figure familiar to all of us, whatever enables us to put ourselves in the role of Ego and then fill in the diagram with actual names of actual people—whatever that learning process has been to each of us—we can be sure that it is adequate to give us the potential to behave so as to maximize our inclusive genetic fitness through nepotism.

PARENTAL BEHAVIOR AND THE
SOCIAL LEARNING MODEL OF NEPOTISM

Parental investment (meaning cytoplasmic and other contributions to the zygote which limit contributions to other zygotes —hence, limit numbers of offspring—see Trivers, 1972; Alexander and Borgia, 1979) is universal among sexual organisms, and parental care, as a form of parental investment, is either universal or nearly so among organisms that are social in any usual sense of the word. Few will disagree with the view that in most social organisms the altruism that is called parental care is adjusted and refined by natural selection so as to maximize the genetic fitness of individual parents through their offspring. This idea dates from Darwin's (1871) remark, quoted earlier, that to increase its rate of offspring production, a parent would have to reduce the amount invested in each individual offspring. To my knowledge, this idea has never been seriously challenged, and as I have shown already, recent tests of either sex ratio theory or reproductive effort alone demonstrate its reasonableness and generality (see also figure 1).

According to an inclusive-fitness model, selection should refine parental altruism as if in response to three hypothetical cost-benefit questions:

1. What is the genetic relationship of the putative offspring to its parents? (Is this juvenile really my own offspring?)

2. What is the need of the offspring? (More properly, what is its ability to translate parental assistance into reproduction?)

3. What alternative uses might a parent make of the resources it can invest in the offspring?

These are the same questions that apply to the analysis of all nepotism. The initially crucial one is that of genetic relatedness. One needs to ask, then, how genetic relatedness might be accurately assessed among organisms in general.

Even for one of the two parents—usually but not always the father—evidence of genetic relatedness is always circumstantial. Females, such as in many mammals, can observe their babies being born and keep them in visual or other contact until unmistakable recognition has been established. Males in forms like seahorses, which first accept unfertilized eggs placed by the female in the male's brood pouch and then fertilize them in the pouch, are in a similar position. But all other assessments of differences in genetic relationship, and thus all behavior appropriate to such differences, are necessarily based on circumstantial evidence. In other words, particular social interactions predict particular genealogical relationships. I may, with some accuracy, assume my siblings to be those individuals who are cared for by the same adult female and male that care for me; a caterpillar may assume as its siblings whoever hatches next to it at about the same time. I assume my offspring to be those juveniles accepted as such by the woman with whom I live. Error is obviously possible in all of these cases.

Many kinds of developmental mechanisms could possibly be imagined to influence discrimination among relatives. It seems to me, though, that nothing more complex or deterministic is required than regular and predictable differences in our learning experiences with different relatives, which lead to regular and predictable differences in our treatment of them. What would have evolved would be our tendencies to behave exactly as learning psychologists already know we do under what they call negative and positive reinforcement—or, more precisely, our tendencies to react as we do to the particular learning schedules which are then labeled as positive and negative because of our reactions to them. I emphasize that it is not necessary, in inclusive-fitness-maximizing, to know who one's kin are, only to behave as though one knows.

There are actually two problems in understanding discriminative nepotism. One is the basis for individuality in attributes that might be used in recognition. The other is how the recognition might take place. The two are not entirely independent. I shall consider the second question first, assuming that sufficient individuality exists to allow recognition of many classes of relatives, but necessarily discussing the probable nature of this individuality along with probable mechanisms of recognition.

The simplest kind of selective action that I have been able to imagine, which might lead to a regularity such as that required for social learning about different relatives, is the accumulation of genes causing us to be positively reinforced according to the number and intensity of physiologically or socially "pleasant" interactions with particular individuals (and we could also be *negatively* reinforced by the opposite). In small groups of genetic relatives these effects alone could cause us to favor closer relatives; and no anthropologist would dispute that humans, during nearly all of their history, have lived in small clans of near and distant genetic relatives. To explain a great deal of sociality, the effects I am suggesting would obviously have to be modified and qualified by many other kinds of learning in different circumstances. To take a radically different example, an utterly and continuously dependent juvenile might gain by being positively reinforced by almost any kind of repeated interactions with an adult—even unpleasant, traumatic ones; this would be adaptive because such a juvenile would usually have no alternative. I believe there is evidence that juveniles are so reinforced—that unusually strong bonds sometimes form between harshly punitive parents and their very young offspring. But this is an unusual case, and considerable generality is implied in the hypothesis that positive reinforcement derives from pleasant interactions. For example, parental teaching about who our relatives are becomes a subset of this hypothesis.

In general, a social learning model of nepotism implies that different combinations of rates, kinds, and timings of social interactions provide an extraordinarily rich source of predictable learning differences which affect the human ability to discriminate relatives and treat them in ways appropriate to inclusive-fitness-maximizing behavior. These effects, moreover, could be extended to include changes in how an individual regards relatives with whom he has no direct interaction. Thus, one may observe interactions of such others with one's own interactants; or respond to accounts of their interactions with one's own interactants, or to accounts of their interactions with others not interacted with, but standing in certain relation to one's self or one's interactants.

To illustrate what I mean by the significance of rates, numbers, and timing of pleasant interactions, and also to show how this approach may possibly lead to new adaptive explanations of ordinary behaviors, I will mention here a familiar interaction between adults and babies. Two human behaviors shared by few other species, and not similarly elaborate in any other species, are tickling and laughing. Tickling and laughter occur together because laughter usually results from pleasant sensations and tickling usually leads to physically pleasant sensations. In particular, adults—more particularly, perhaps, parents—tend to tickle babies, causing them to laugh. A characteristic behavior when babies are tickled is for the tickler to place his or her face directly in front of that of the child and repeatedly draw closer as he or she tickles. Thus, the baby is repeatedly subjected to the sight of the smiling face of the tickler passing through the plane of its visual focus as the pleasant sensation of tickling is experienced, and that this is actually occurring is communicated to the tickler by the baby's laughter. This kind of rapid repetition of what is evidently a maximally pleasurable experience between two individuals positioned so as to maximize the probability of recognition may lead to exactly the kind of accumulation of learning experience that I would postulate to underlie our abilities and tendencies to learn who are our relatives and friends, what to expect from them, and how to behave toward them. I think it is even reasonable to hypothesize that tickling and laughter may have evolved as parts of a special mechanism of socialization in humans, socially the most complex of all species.

I would argue that to understand, predict, and adjust human social behavior in modern environments will require description and analysis of all possible and probable combinations of relevant social learning situations explicitly in the light of their effects on inclusive-fitness-maximizing. I suggest that this enormous task may well become a central problem in the social, developmental, and learning psychologies of the future. Such an approach is already hinted in recent analyses of social transactions, manipulations, coalitions, power interactions, and reciprocal interactions and networks, particularly as discussed by Boissevain (1974), Blau (1965), Homans (1974), and Hatfield et al. (1979). The flaw of these studies to date, as I have already suggested, is their universal failure to distinguish interactions that are nepotistic and surrogate-nepotistic (i.e., nepotistic in historical terms) from those that are reciprocal and not nepotistic (see also figures 5 and 6). Without an understanding of the centrality of nepotism, complete and satisfying explanations, particularly of what most of the above authors term "deep" or "intimate" interactions, have not been possible; elsewhere I have inferred that such interactions are usually nepotistic (Alexander, 1979c).

Models of Proximate Mechanisms of Nepotism Other than Social Learning

The "Genetic" Model
One might at first suppose that the appropriate dispensing of discriminative nepotism could be, or even has to be, based on

the appearance of mutant genes which enable their bearers to recognize and respond to their own effects in other individuals. As Hamilton (1964) noted, an appropriate genetic unit would have to, first, have an effect on the phenotype; second, cause its bearer to be able to recognize the same effect when it occurred in others; and, third, also cause its bearer to take the appropriate social action. These requirements of gene action are probably unreasonably complex to expect in a single mutant, but they could occur in a group of tightly linked genes behaving like a supergene, where such linkage would tend to benefit genes with each of the three kinds of effects. This model is weakened, however, by the fact that genes which only gave the phenotypic effect, without any altruistic tendencies, would often spread faster than those doing both.

More important in judging the fate of any mutant or supergene like that described above is the likelihood that it would act solely on the basis of its own presence in the genome of the potential recipient of its bearer's altruism, irrespective of the presence of the alleles occurring at other loci in its own genome. Should this occur, the action of such a mutant would usually be deleterious to all other genes in the genome and the extraordinary organization of the genome could not be sustained if such genes became prevalent. This is a particularly important point, since this hypothetical mechanism does not restrict nepotism to relatives by descent; it could operate between any two individuals with the relevant genetic unit in common, and this would increase the likelihood that other genetic units would not be present in the genotypes of both the helper and the helped. Any gene mutating so as to suppress such an "outlaw" effect by a subgenotypic unit, even partially, would thereby help itself, and in any large genome the mutational probabilities of suppression of single genes or other small genetic units, as opposed to the probability of sustaining outlawry, would be enormously high (Alexander, 1977a; Alexander and Borgia, 1978).

Seger (1976) has provided a model that is interesting to consider in light of intragenomic conflicts of interest. He notes that organisms in highly inbred populations will not only tend to have a higher proportion of homozygous loci than those in outbred populations but will also tend to have less genetic conflict of interest with individuals with which they closely associate and would normally compete for resources. If some mechanism existed whereby homozygosity could be assessed, selection should result in individuals that are least competitive toward associates when their own genomes are most homozygous and most competitive toward associates when their genomes are most heterozygous. This effect could feasibly occur as a result of the collective action of individual alleles somehow producing effects by which they could recognize their own copies at the same locus and modify effects on their bearer's competitive activities accordingly. Assuming that such outcomes are possible, whether or not they represent a significant evolutionary force would seem to depend upon population structure. If homozygosity occurred in many different degrees and fluctuated erratically, this kind of activity by individual genes would often be deleterious to genes at other loci with differing probabilities of appearance in the genomes of associates of their bearers; therefore, such effects would likely be suppressed (Alexander and Borgia, 1978). If, at the other extreme, the variation was usually between the two states of genome-wide homozygosity and genome-wide heterozygosity, then the interests of genes at individual loci within genomes would be more likely to correspond. One might also imagine that individual genes could evolve to "read" the extent of homozygosity at many or most loci. Such genes could produce effects advantageous simultaneously to themselves and most of the genome if sufficient consistency existed between their own likelihood of representation in competing genomes and that of all the loci read for homozygosity.

In other words, when looking at the redistribution of resources among relatives, or at all forms of parental and other nepotistic effort (as contrasted with somatic effort—Alexander and Borgia, 1979), it is misleading to ask simply whether a single gene will spread itself by the effects of its action on its bearer. In all such cases there is a considerable potential for a single gene to affect deleteriously the reproduction of other genes with which it will necessarily coexist every now and then in the same genome. To consider genes singly in such circumstances is analogous to supposing that individuals in societies can pursue their own self-interests without opposition, when the interests of others suffer. That neither individuals nor genes can be so understood is a simple, yet often neglected point.

Determining Degree of Relatedness by Phenotype: Nonrandom Sampling
Parents, in ordinary sexual species, are truly 50 percent like their actual offspring in genes identical by immediate descent, if we ignore the slight asymmetries caused by the sex chromosomes. But the genetic similarities among other relatives, because of the uncertainties of meiosis, are only averages. Full siblings only average 50 percent alike, and any particular pair of full siblings is likely actually to share either more or fewer than half their genes. It seems plausible, therefore, that the same kind of phenotypic judgments sometimes used to assess whether or not an individual is one's own offspring could also be used to judge the actual proportion of overlap with individual relatives such as full siblings. Not so. Any genetic unit that contributed to a tendency to assess the actual amount of genetic overlap between its bearer and a full sibling would extinguish itself unless it used only its own expressions to make the assessment, returning us to the first ("genetic") model above. Genes with such ability would again not only have to be extraordinarily complex and specific in their action, but would tend to break up the genome, thus removing the empirical reason for supposing that the individual has been an important focus of evolution by natural selection.

Determining Degree of Relatedness by Phenotype: Random Sampling
Suppose that a mutant gene gave to its bearer the ability to sample randomly the phenotypic effects of all the genes of other individuals and then to compare these effects with the bearer's own phenotype. An individual with this capability could as-

sume with some accuracy that its genetic overlap with the other individuals was proportional to the overlap in the random sample. A capacity to sample the effects of the entire genome, however, locus by locus, is not even remotely likely. If the sample were drawn from only a few loci, even though randomly, phenotypic homogeneity for those loci or the appearance of "mimics" lacking altruistic effects on the phenotype would soon render the capacity useless. My arguments regarding these two models obviously mean that I reject the models proposed by Barash et al. (1978).

Acceptance or Rejection by Comparing Phenotypes

Heritable variations in phenotypic traits may be used in some species to assess the likelihood that a particular individual is or is not a particular relative. On occasion they are so used in our own species. Thus, a man may doubt that an offspring is his own if the child's hair color, eye color, or other attributes deviate too strikingly from his own (or too closely resemble those of another). This mechanism requires genes that collectively give their bearers the three abilities listed by Hamilton (see above), but it does not require that the ability to recognize phenotypic attributes be conferred by the same genes causing the attributes, or that such abilities be used to make quantitative (as opposed to qualitative) judgments. Most importantly, it does not specify how the appropriate response to a particular phenotype is established; the judgment probably requires previous learning about one's own phenotype (or those of others) as well as about the nature or significance of the particular relationship being accepted or rejected.

It may be supposed that a close parallel exists between the mechanisms of nepotism and those responsible for the immune reactions by which organisms distinguish their own molecular forms from those of other species, or by which they discriminate among many different disease organisms and parasites and produce antibodies specific to each. The parallel, however, may not be so close. The part of the organism's genome that participates in creating antibodies against foreign antigens would normally contribute to the success of the entire genome; it could not, as in nepotism, help itself or its copies differentially. This is true because the immune response in natural situations is generally a negative response toward potentially harmful members of a different species (the matching of organs for transplants between individuals of the same species is an evolutionary novelty and irrelevant for these particular considerations). Consequently, immune responses can involve even small portions of the genome without threatening its integrity, and different portions of the genome can feasibly be involved in responses to different foreign antigens.

In regard to proximate mechanisms, immune responses among sexual organisms seem more comparable to behavioral responses toward predators, parasites, prey, or symbionts than to beneficence toward relatives in one's own species. Responses between mammalian mothers and their fetal offspring are an exception, but they are still not a good model for the origin of discriminative behavioral responses during nepotism. The mammalian mother has probably evolved to accept whatever embryo develops inside her, since she could normally carry only her own offspring; and she has probably evolved various ways of avoiding deleterious immunological effects on her own offspring. Thus, hormones specific to pregnancy reduce immunological reactivity, and paternal-maternal grafts are rejected more slowly during pregnancy (Johansen, 1977). The so-called "immunological incompetence" of mammalian embryos (and unhatched embryos of other forms), whereby they do not treat foreign proteins as such, suggests that embryos have evolved to be immunologically tolerant of foreign products from their mothers, to which they are liable to be exposed during fetal life (or while in the egg) (Johansen, 1977; Cooper, 1976). The evolutionary (selective) background of the particular kinds of incompatibility that do exist now and then between mothers and their fetal offspring (such as the rh factor) appears to be largely an unsolved problem; some such reactions, however, probably result from the mixing of previously noninterbreeding populations.

It is also useful to compare the probable mechanisms of nepotism with those that enable insects or other animals to recognize the sexual signals or other phenotypic attributes of members of their own species without any prior experience in that context. The most extensively analyzed case is probably that of the songs of male crickets, which attract sexually responsive females (e.g., Alexander, 1969). The crucial part of the song pattern depends upon the firing rate of a central nervous system pacemaker, which is almost certainly dependent on rates of membrane polarization, in turn dependent on the structure of the proteins in the membrane. Thus, the song pattern may be recognized by a trait that is only the two steps of transcription and translation from the action of the genes. Such a brief stable ontogeny for complex sexual signals leads one to wonder if the genes themselves (in the somatic cells) have been treated by natural selection as an unvarying aspect of the environment of the phenotype from which to guide the development of behavior; this would approach as closely as I can imagine what might be termed "innate" or "genetically determined" behavior.

Sexual recognition signals, however, also differ from discriminative nepotism in the sense that the genes involved in either producing a signal or responding to it could not help themselves at the expense of the rest of the genome. There is evidence that the genes responsible for the patterning of a signal may also underlie the ability to respond to that same signal (Hoy, Hahn, and Paul, 1977). A few genes, identical in all members of a breeding population, could feasibly be responsible not only for the signal produced by each individual but for the response to the same signal produced by the same genes in other individuals, and yet there would be no intragenomic conflict, as in nepotism. Something similar is true for outbreeding mechanisms, such as those apparently existing in *Drosophila* (Averhoff and Richardson, 1976), since no gene would gain from inbreeding if it were thereby placed into a genome containing other genes that when homozygous produced deleterious traits.

There is much more to say on this topic. I believe that a

fundamental ontogenetic difference may exist between the sexual and social signals of insects and lower vertebrates and those of most birds and mammals. This difference may account for the incredible stereotypy or species-wide monomorphy of insect and amphibian acoustical signals through an evolutionary convergence of the CNS template for the song pattern and the response to it on population- and species-wide bases. It may also indirectly resolve the paradox of the simultaneous existence of (1) species-specificity in reproductively isolated forms which breed together in the same times and places, (2) an absence of consistent differences between such forms that do not breed in the same times and places, and (3) no clear evidence of character displacement (exaggerated divergence as a result of competition or interference) in forms that overlap across only parts of their ranges (Alexander, ms.). In other words, the selective pressure for identity in sexual signals in insects, coupled with the pressure for divergence in the areas of overlap, could cause rapid spread of the displacement, right to the border of the species range in many cases, even far outside the areas of overlap. Several predictions result. Thus, exaggerated character displacement in areas of overlap would rarely be observed. Moreover, it would be likely to be represented by stepped clines (areas of change) with the steps representing range discontinuities (or barriers to dispersal). Character displacement should also be more common in birds and mammals that acquire their sexual signals by social learning and benefit from some individuality. Finally, "ecological" character displacement (for example, resulting from competition for food rather than for mates) should be more common than "reproductive" character displacement (that involving sexual signals). Very likely, all of these things are true (e.g., Grant, 1972; Alexander, 1967; Walker, 1974).

The important point here is that in the absence of learning one expects social responses, and the phenotypes which are their objects, to be singular and uniform, never individualized, among all the members of a population or species. I would regard the onus of proof to be on the investigator who argues that among birds or mammals no social learning opportunities have been involved in nepotistic or reciprocal interactions in which individuals are treated differently, or who applies to such behavior adjectives like "innate," "instinctive," "inborn," "genetically determined," or "unlearned." These terms, when used to describe social behavior, typically do not refer to any developmental mechanism. They are instead negative with respect to models or mechanisms, describing only what the author believes is *not* involved. They tend to involve an unjustifiable leap from an absence of information about ontogeny, or from a knowledge that differences in the behavior involved correlate with genetic differences, to the implication that only genes underlie the behavior—that there is no learning, no modifiability, indeed, no ontogeny at all. Wilson (1975) demonstrates the uselessness of the above labels in an unusual effort to avoid their negative connotations. He says that "an instinct, or innate behavior pattern, is a behavior pattern that either is subject to relatively little modification in the lifetime of the organism, or varies very little throughout the population, or (preferably) both" (p. 26) and that "learning may or may not be involved in the development of instinctive behavior; the important point is that the behavior develops toward a narrow, predictable end product" (p. 587). This curious definition allows him (1978, p. 38) to refer to incest avoidance as follows: "To put the idea in its starkest form, one that acknowledges but temporarily bypasses the intervening developmental process, human beings are guided by an instinct based on genes." How the intervening developmental process is acknowledged by this statement is obscure. What is added beyond the knowledge that sex is avoided with those with whom we associate very closely as juveniles (Wolf, 1966, 1968; Shepfer, 1971, 1978; Shapiro, 1958) is also obscure. That incest avoidance is socially learned is *made* obscure. Moreover, by Wilson's definition almost any behavior might be called an instinct—such as how the word instinct is spelled or pronounced. And the implication of no ontogeny, no modifiability, no learning remains: if not otherwise this is surely illustrated by the acrimonious, unnecessary, and largely unprofitable controversies that followed in the wake of Wilson's 1975 volume (e.g., see Caplan, 1978). It seems to me that a word is needed for behaviors with cryptic or hidden ontogenies so that they will not be labeled as "genetic," "instinctive," or in other misleading fashions. Until such a term appears it is only appropriate that authors be called upon to describe what their observations or experiments eliminate as possible ontogenetic or experiential backgrounds and discouraged from labeling a behavior "innate" simply because its ontogeny is unknown.

So we are returned to social learning as the chief means by which the genes in the individual's genome are able to realize their *common* interests in nepotism. We are not led to some kind of unacceptable "genetic determinism." Indeed, the rather general possibility among mammals of inducing adoptions of nonrelatives, by manipulating the social circumstances, is alone sufficient to indicate that nepotism is likely based on the social interactions of the involved individuals. These social interactions will tend to benefit all of the genes of an individual equally when they cause individuals to act on the basis that other individuals belong to one or another *class* of relative (sibling, niece, cousin, etc.). For each class of relative all genes in the genome of Ego share a certain probability of being present (figure 4). Thus, a social learning mechanism allows for the remarkable unity of the genotype and the phenotype produced by it.

Social Learning and Individuality

The premium on individuality in human social interactions raises questions about phenotypic variations in traits like facial features, ear shape, hair and eye colors, and balding patterns. Some of the variations in these traits are surprisingly closely linked to genetic variations and have also been judged to be trivial in their adaptive significance. It is of course possible that such variations, though we may frequently use them to identify individuals, are linked, as with blood types, to other traits of more evident adaptive significance (e.g., see Ford, 1971). Another possibility, however, is that both the phenotypic variations and their close linkage to genetic variations are a consequence of selection favoring individually recognizable combinations of traits, and traits that can characterize clans or groups of related individuals. Thus, one may learn from observing either his own traits, or the traits of relatives who become known for other reasons, that particular sets of phenotypic attributes tend to characterize one's genetic relatives. Appropriate responses to kin, then, can be developed from a complex coupling of social learning and the use of genetically determined differences in phenotypic attributes. I suspect that essentially all human sociality among kin will be discovered to be based on such combinations of responses. It is not surprising that their effectiveness should be enhanced by cultural phenomena like language, passwords, clothing, ornaments, and secret gestures or handshakes.

This argument casts interesting light on the four models suggested above, for it suggests that continual phenotypic comparisons of known (socially learned) relatives and individuals of unknown relatedness may be responsible for much social learning relevant to inclusive-fitness-maximizing behavior through both nepotism and genetic outbreeding. Thus, while it may not be possible to sample gene effects randomly on the phenotypes of interactants so as to ascertain probability of relatedness, it may nevertheless be entirely feasible to gain from comparing points of similarity between relatives of known and unknown degree. One reason is that phenotypic differences owing to gene differences, and used in such comparisons, are unlikely to be either fixed or obfuscated by mimics as they would be if they were being used solely in quantitative assessments of the types discussed in connection with the models not involving social learning.

Special Cases of Social Learning: Falling in Love

The above arguments about the effects of social learning upon nepotism do not account for all cases of beneficence;

same genes—or the same supergene—would be equally valuable to everyone if the useful kinds of information and the learning situations were similar for all individuals. Genes leading to patterns of nepotism that maximize fitness should, in fact, become fixed in the population, even if their action were extremely indirect through a wide variety of learning processes. In the extreme, one could imagine the fixation of a single supergene in all humans which gave to every individual the ability and tendency, in normal environments, to discriminate relatedness and needs of relatives optimally for his own genetic interests. Even if this is an unlikely possibility because of cultural variations among societies, and temporal changes within societies, it is a useful postulate for thinking about effects of selection on inclusive-fitness-maximizing behavior.

The paradox in the above scheme is that it leads to identity among individuals in precisely the genes responsible for abilities and tendencies to discriminate among relatives, and thus also responsible for competition among individuals (because two individuals almost never have exactly the same set of relatives—an exception being monozygotic twins). One supposes that, with fixation of the above supergene, tendencies would spread to ignore genetic differences and treat everyone equally because all would share the genes actually responsible for discrimination. No mutant leading to such behavior, however, could invade the system just described. Each new mutant on a supergene for nepotism would reintroduce a significance for discriminatory behavior on the basis of genes identical by immediate descent among interacting relatives.

It would do this because, at the start, its own distribution would be based on immediate descent. This means, paradoxically, that mutants causing any departure whatsoever from the kind of inclusive-fitness-maximizing behavior just described would, by this particular effect on their phenotypes, render themselves inferior alternatives to the established alleles with which they competed. The only way in which a supergene for nepotism of the sort described could be altered or replaced would be if changes in the cultural environment diminished the tendency of the old supergene to lead to inclusive-fitness-maximizing, so that a mutant could improve this tendency.

Washburn (1978) has argued that as few as 1 percent of the genes of humans may vary among "non-relatives," too few to cause nepotism to be evolutionarily significant. The data Washburn cites, however, refer not to genes but to genetic "information." A great deal of the DNA could be identical in two organisms which shared not a single allele, defined as a unit of inheritance and recombination. The above argument shows, however, that even if 1 percent is an accurate figure selection could maintain discriminative nepotism. But the variations in success of transferring organs like kidneys between relatives and nonrelatives, and the greater extent of immune-system matches in the former, suggest rather that variations in the number of genes held in common by different relatives are in no way trivial, and that their proportions vary as the diagram of relatives in figure 4 indicates. Thus, Barnes et al. (1968) report the following percentages of successful kidney transplants: monozygotic twins, 80 percent; parent-offspring, 68 percent; siblings, 62 percent; other relatives, 44 percent; nonrelatives, 39 percent (cadavers), 27 percent (living). The last figures, moreover, are undoubtedly elevated because of careful tissue matching between potential recipients and donors.

These arguments show why Hamilton (1964) was correct to focus his analysis upon relatedness in genes identical by immediate descent, and they also suggest how to deal with questions of genetic overlap from inbreeding and convergence owing to parallel or convergent selection. One needs only to consider the fates of mutants affecting nepotistic behavior. Such mutants represent the means by which the altruism of nepotism generates, increases, and becomes directed with precision. The successive waves of such mutants will always maximize their own spread by treating relatives as if their own likelihood of occurring in the relative depends upon the proportion of genes identical by immediate descent. This is because each new mutant will at first indeed tend to be present in just those proportions: for this reason better odds will not occur.

The single environment in which all that I have just said can become irrelevant, of course, is that in which the interactants have become consciously aware of this aspect of their natural history. Perhaps that is the most important point of this book.

CONSCIOUSNESS, FORESIGHT, AND FREE WILL

I have argued so far that culture is a product of the efforts of all of the individuals of history, in the different environments of history, to maximize their separate inclusive fitnesses, and I have described mechanisms, based on ordinary learning phenomena, which could account for such efforts. I have suggested that the capacity for culture is really the capacity to use a wide array of experiences to adjust one's personal outcomes in the interest of maximizing reproductive success. Now I would like to relate this view of human behavior to certain distinctive phenomena of human existence which are often regarded as inaccessible from biological theory.

Consciousness is a system by which we are in some sense aware of ourselves and our relationship to others and the rest of the world. For this reason I regard it as most likely evolved in the context of success in social matters. My own introspection, and my observation of others, suggest to me that how others see us is absolutely crucial to social success, however success is defined. It appears to me that consciousness is largely a system for testing the question of how others see us and adjusting our image in our own self-interest—that it is a vehicle for, as Robert Burns put it, seeing ourselves as others see us.

A central aspect of consciousness is the ability to look ahead, the capability that we call "foresight." It is the ability to plan, and in social terms the ability to outline a scenario of what is likely going to happen, or what might happen, in social interactions that have not yet taken place. The significance of this ability is obvious. It is a system whereby we improve our chances of doing those things that will represent our own best interests. We consider what will happen if—if we do this or if we do that; and we try to judge what others with whom we expect to interact will do, under each of several circumstances, and to figure out how to get them to behave as we would like them to. We draw up alternatives and consider them, one by one. It is especially significant that we do this so prominently in connection with social interactions, for no other life situations are quite so uncertain, or so important to plan out ahead of time. The reason is that those with whom we interact are also capable of scenario-building, and so their behavior will be tuned to their own particular interests and their own efforts to anticipate our responses. No feature of the environment is quite so difficult to figure out as what to expect from other social beings with whom we must interact, each of whom is attempting with all of the capabilities he can muster to adjust the outcomes of our interactions with him to his own advantage, rather than to ours, when our interests differ.

In the course of using our consciousness and foresight to build scenarios and plan our social interactions, we visualize alternatives and test them one by one. We see these different alternatives as available to us if we choose to use them, and in some sense they are—or at least some of them are. I believe that it is our ability to visualize alternatives, particularly in connection with social interactions, that represents the basis for the concept of free will. We see points of decision ahead of us because we have used all of the information we can muster from the past and present to build scenarios about our immediate futures, and I suggest that "free will" is our apparent ability to choose and act upon whichever of these decisions seem most useful or appropriate, and our insistence upon the idea that such choices are our own.

In the sense that I have just described it, free will is not incompatible with the notion of an evolved tendency to maximize inclusive fitness. Indeed, I believe that it is easily seen as an outcome of selection leading to inclusive-fitness-maximizing through learning in the particular kinds of social situations in which humans have evolved. Moreover, it is easy to understand why so much of life is theater for us: theater, in all of its guises, is perhaps by definition the richest and most condensed of all cultural contributions to our patterns of scenario-building through consciousness and foresight.

What Is Conscience?

If humans are evolved to maximize their inclusive fitness, which will differ for each individual (excepting identical twins), and if free will means the right to make one's own cost-benefit decisions about inclusive-fitness-maximizing, then what is conscience? I suggest that conscience is the still small voice that tells us how far we can go without incurring intolerable risks. It tells us not to avoid cheating, but how we can cheat socially without getting caught. Conscience plagues us even when we do not get caught, and sometimes causes us to confess when it may seem that there is no longer any risk, because by experiencing such consequences we teach ourselves not to do again under the same circumstances those things that turned out to be too risky. I also suggest that most of the time confessions actually occur when there is a risk of being detected. This kind of self-teaching is only a shade removed from the deliberate self-conditioning used by psychologists to help individuals refrain from self-destructive behaviors, from drug addiction to sexual and social crimes. Reviewing the risks we have suffered in a breach of societal standards is only a particular aspect of the scenario-building that I have described as the central function of self-awareness and consciousness—one associated with actions, not of someone else, but of ourselves, that we do not wish to repeat or that we want to test in regard to the possibility of repetition. Perhaps our conscience will stop hurting, or perhaps it will hurt more; our future behavior in regard to the same event may hinge on such questions.

Finally, I suggest that conscience appears to be a more powerful force in "respectable" people because they are those who have decided to maximize their inclusive fitnesses by following the rules (see chapter 4), and once that decision has been made they will be better off knowing how to accomplish their goal.

Self-Deception

It is a remarkable fact that humans have not only failed, throughout history, to acquire any understanding that they have evolved to maximize reproduction, but that even today they deny the possibility vehemently. Even if the entire idea turns out to be wrong, which seems exceedingly unlikely at this point, we can still marvel at the hostility it engenders. I suggest that this attitude derives partly from the importance we attach to what our associates think of us. No one is less attractive as an associate than someone who is known to be utterly selfish, with only his own interests at heart, or is a known liar—one who deceives deliberately and in the circumstances in which the gain to himself is the greatest. Additionally, whatever pictures of ourselves may have generated in our minds before now, they had to be constructed without knowledge of the subgenomic units underlying it all.

But the view of human history developed so far in this book suggests that humans behave as if they are concerned with their own genetic interests, and that they are also masters at deceiving others. I suggest that the separateness of our individual self-interests, and the conflicts among us that derive from this separateness, have created a social milieu in which, paradoxically, the only way we can actually maximize our own self-interest and deceive successfully is by continually denying—at least in certain social arenas—that we are doing such things. By conveying the impression that we do not intend to deceive, and that we are in fact altruistic and have the interests of others at heart, we actually advance our own (evolutionary) self-interest. I believe as a consequence that our general cleverness at creating deceptions and detecting them has made it next to impossible for individuals to benefit from deliberate deception in ordinary social situations, because of the likelihood of detection and exposure, and, possibly, severe punishment. The result, I believe, is that in our social scenario-building we have evolved to deceive even ourselves about our true motives.

Perhaps the most remarkable case, which seems to me to

support strongly the argument I am presenting, is the evident concealment of ovulation by human females not only from others around them, including their mates, but from themselves (Alexander and Noonan, 1979). Ovulation is the presentation of an egg for fertilization, so it is an important event in a female's reproduction. Many women know approximately when they ovulate, and some may know rather precisely. Nevertheless, this knowledge is sometimes available only because of certain kinds of recently acquired medical knowledge and technology, and it sometimes becomes conscious only because of special concerns of a medical nature. The human female clearly has not evolved to keep knowledge of ovulation in the center of her consciousness, where, in view of its importance, we might have expected it to be. At first, it seems that even if it were advantageous to the human female to conceal ovulation from all those around her, including her mate, she herself would still profit from knowing precisely when it occurred and being keenly aware of it. But this kind of knowledge would entail continual conscious and deliberate deception of her mate and others; perhaps such deception is contrary to our basic way of operating socially, and it would be too difficult, and too discordant with the other aspects of sociality, for it to be maintained.

I have used the example of concealed ovulation—sometimes discussed as "continuous sexual receptivity"—because it is so profoundly important in understanding humans, and because it involves a physiological event so central to reproductive success. Alexander and Noonan (1979) have argued that concealment of ovulation correlates with a massive increase in male parental care in humans, and that this increase in turn has led to the dramatic neotenic trend (toward juvenile traits in older individuals) that has for so long puzzled the students of human evolution (see pp. 209–16).

Eventually we ought to be able to disentangle and make sense of an enormous number of social interactions with particular historical significance within the context of self-deception. I assume that such insights will help free us from insecurities, uneasiness, and sources of tension and unhappiness which are all too often cryptic and unconscious. I assume that such insights will disencumber our efforts to pursue social interactions disinterestedly. I assume, both optimistically and pessimistically, that the only answer to our imperfect understanding of ourselves is a less imperfect one. Partly this means I believe that those who suggest that we should stop evolutionary analyses, because they can only lead to oversimplification and error, are wrong.

In considering the question of self-deception it is not trivial to recall, again, that the genes and their mission, before the advent of modern science and technology, remained outside human knowledge. So we could not know what we are really all about. In the face of this crucial ignorance our efforts to maximize inclusive fitness may often have led us into ways of achieving our ends that look like self-deception but are not. If the real reasons for treating first cousins differently from second cousins are inaccessible to us, we might expect that the reasons devised or accepted by us might be nonsensical, or illogical, and thus that some degree of blind faith or even self-deception might be involved in their acceptance. Indeed, this problem may turn out to represent a major source of so-called arbitrariness in culture: arbitrariness of reasons given and even widely or universally accepted, but not arbitrariness of action with regard to genetic reproduction.

Not all actions are described or accepted for arbitrary or false reasons. Sometimes it is remarkable how close we came, not knowing about genes. We know which among many relatives are "closer," hence almost "know" degrees of relatedness, and we speak of "blood" relatives versus "relatives" by marriage. In an example discussed later the reasons given by the natives for certain kinds of cousin marriages being preferred are closer to the truth than either the natives themselves or the anthropologists studying their culture—each operating without knowledge of genes—could possibly understand.

The challenge of Darwinism is to find out what our genes

have been up to and to make that knowledge widely available as a part of the environment in which each of us develops and lives so that we can decide for ourselves, quite deliberately, to what extent we wish to go along. I think we can be certain that our genes have not evolved to make the knowledge of their strategies conscious. If we accomplish such understanding it will be an accidental effect, not an evolved function of the genes. In no way, however, can our freedom of choice be restricted by knowledge of the effects of culture upon the directions of strivings of which we formerly were unaware. To the contrary, I can imagine no single kind of information that would so dramatically enhance freedom of choice. I like the way it was put by Rosenfeld (1977):

> . . . the individual who militantly seeks to have the quest for knowledge brought to a halt is often the same individual who is outraged by the sociobiological suggestion that we are more controlled by our genes than we realize. We *are* more controlled by our genes than we have realized [this assertion is eminently reasonable, since only a few decades ago we were entirely unaware of the existence of genes]; therefore, the more we discover about the mechanisms of genetic control, the better equipped we will be to escape those controls through our enhanced awareness, to transcend them so that we may, for the first time in our history, work for ourselves instead of for our genes, exercise truly free will and free choice, give free reign to our minds and spirits, attain something close to our full humanhood. [Pp. 19–20]

It is fashionable nowadays to believe that the ultimate contribution of the biologist to an understanding of human behavior, if not to *predict* its expressions, is at least to describe its theoretical limits—the limits of what we will be able to achieve in designing and developing the structure of society. In subscribing to Rosenfeld's view I am clearly not so foolhardy as to make such an attempt. Who can say, I would rather ask, what humans —who have evolved to maximize inclusive fitness in the environments of history, and who began long ago to probe the extraterrestrial as well as the subatomic—will be able to accomplish in a new social environment of enlightened self-reference that includes knowledge of our evolutionary background and the mission of our genes? In the first place the people who change their attitudes upon learning from modern natural history where they have come from, what they are, and why, themselves become and create the environment in which any presumed limitations from biology will be realized. Who can presume to know every direction and degree of such changes in attitude and all of their effects?

Second, the magnitudes and directions of change in our environment owing to advances in technology are not entirely predictable: who could have predicted, only a few decades ago, the nature and social significance of devices by which conception could easily be controlled?

Third, I have a deeply ingrained hesitancy at identifying what is "best" for anyone but myself. Even for my closest and most dependent relatives (including my own infant offspring), unless I am confident of a correctable difference in our information about the event in question, I want to know their views before I carry out an act or lay down a rule. I prefer to live in a society in which individuals incline in this direction rather than any other.

Finally, designed societies are by definition restrictions upon diversity, therefore upon opportunities to judge alternatives. The more effective the design (e.g., the greater its effective projection in time), the smaller the group of designers, and the more all-encompassing the rules, the greater is the resulting narrowing of diversity. I see all of these as cogent reasons for leaving aside the question of grand changes in the design of society, especially in the immediate wake of an alteration in our view of ourselves as profound as that we are currently undergoing.

Edward O. Wilson's *On Human Nature* (1978) is important as the first book devoted to a philosophical discussion of humans, yet incorporating to a large extent the modern view of evolution. Nevertheless, it is puzzling that Wilson calls for a retention of human genetic diversity, yet seems to believe that better understanding of our natural history will lead inevitably to a "more enduring" moral code (expressed in the singular) and a "narrowing" of our social "trajectory." There is an implication of desirability—or ethical opinion—about his statements on these topics. He says that genetic diversity should be maintained because we need the intellectual geniuses produced by recombination. He does not, however, describe the virtues of inducing singularity in moral codes or social trajectories. It seems to me paradoxical to call simultaneously for genetic diversity and behavioral conformity—even, or perhaps especially, in regard to opinions about social, moral, or ethical matters. I am content to leave what "ought to be" to the entire collective of people willing to think on such problems and participate in their solution, and to view the contribution of biology as one of enlightenment about natural history—what is and was, and perhaps what will be *if*—rather than what ought to be. I shall have more to say on this topic later.

3

Natural Selection
and Patterns of Human Sociality

Introduction

I have argued that pursuing the logic of Darwinian evolution leads us to the conclusions that all life is derived from organic evolution, that evolution is guided chiefly by natural selection, and that selection is almost never effective above the individual level. The consequence is that life is appropriately viewed as comprising individual organisms whose lifetimes have been designed by evolution to maximize the likelihood of persistence of their genes. Whatever else lifetimes may be, by this theory that something else should be an evolutionary accident or a consequence of novelty in the environment.

In this view of life, human culture is seen as an environment of symbols, rules, traditions, and other products of human inventiveness, used and modified by the individuals of one generation, and both available to and imposed upon the individuals of the next. Such an individually oriented view of human sociality is actually not unique to modern biology. Masters (ms.) has traced its origins in what he called the "hedonistic individualism of the Sophists" from the Greeks forward; such a view has

nd Wilson, 1978; Hartung, 1976; van den Berghe and
1977; Weinrich, 1977; Wilson, 1975, 1978. Wilson,
nd Caplan, 1978, are especially rich in references not
my discussions).

THE HUMAN NETWORK OF KIN

evolutionary model predicts, then, that viscous social
s like human societies should be extraordinarily complex
rks of nepotistic interactions. It is of paramount impor-
that this is true for humans. At the base of every known
n society there is a kinship system in which genetic rela-
ips are an essential ingredient. Every human from birth
bedded in such a network. Indeed, social anthropologists
always regarded it as their central role to examine human
ip systems and analyze their structure and functions. Ear-
suggested that social psychology would someday recog-
as a central task the explanation of all of the forms and
exts of social learning as *mechanisms* of inclusive-fitness-
imizing. Here I suggest that social anthropology will some-
recognize that one of its central tasks is explaining varia-
s in social behavior and in the patterns of culture as *outcomes*
nclusive-fitness-maximizing.

When humans began to live continuously in social groups
hin which they could distinguish relatives of differing de-
e, two new selective forces were added to their social life, as
mpared to that of most other species. First, numerous and
ried extrafamilial relatives became reproductive resources as
th sources and targets of nepotism, and, second, genetic out-
eeding beyond the limits of the nuclear family became possi-
e because information about the differing degrees of related-
ss of extrafamilial relatives became potentially available to
very individual.

Assuming the situation described in the above paragraphs to
e the case, and taking a modern view of Darwinian selection,
everal general predictions about human nepotism are evident,
nd numerous subsidiary predictions emerge. For example, we

also developed independently among modern social exchange
and network theorists in social psychology and anthropology
(e.g., Boissevain, 1975; Barth, 1967). Late in his life the social
anthropologist George Peter Murdock (1972) came to this view:
"It now seems to me distressingly obvious that culture, social
systems, and all comparable supraindividual concepts . . . are
illusory conceptual abstractions inferred from observations of
the very real phenomena of individuals interacting with one
another and with their natural environments . . . culture and
social structure are actually mere epiphenomena—derivative
products of the social interaction of pluralities of individuals"
(p. 19). Curiously, the significance of Murdock's change of view
was lost upon anthropologists; I have seen it quoted only once
(Sahlins, 1976a), and then to illustrate what Sahlins saw as
Murdock's error.

The conclusion that sociality is a consequence of individuals
pursuing their genetic interests is discomforting not only to
those who have consistently searched for function at group
levels in human affairs, but equally so to community ecologists
and other biologists whose research has been similarly directed.
The implication in both cases is that the evolutionary significance of function is not to be found at high levels. This is a
distressing outcome to those who have sought particular kinds
of simplifying generalizations to account for the complexity of
organization of life. Thus, it seems incredible to suggest that all
principles regarding ecosystems, communities, ecological succession, interspecific interactions, and the patterns of culture
must derive from natural selection focused at and below the
individual level. I think, however, that for nonhuman
phenomena we are already forced to this conclusion and that it
is at least the best hypothesis for human sociality and culture.

I have also argued that cultural evolution differs from genetic
evolution because it incorporates a feedback between the needs
of the organisms and the sources of novelty in behavior. The
causes of change and the causes of the spread and persistence
of change are not independent. The mutations of culture—
discovery, invention, and planning—are directional, and their

heritability is adjustable on the basis of their suitability. To understand directions and rates of cultural change, then, we must concentrate not on heritability as such, but on who or what causes and keeps the mutations of culture, and on what basis.

I have suggested that culture is kept "on track" by the interests of individuals, and not by the interests of populations or species except in the case that the interests of the individuals comprising them are shared; that it represents the collective effect of all individuals trying as best they can to match their perceptions of their own best interests—whether these are conscious or somehow unconscious. The nature of their perceptions is the ingredient omitted from the formulations of cultural determinists. I have suggested further that human individuals, like other organisms, have evolved to interpret their best interests (not necessarily consciously) in terms of reproductive maximization. It should follow that even the environmental novelties introduced by culture will not often be irrelevant to reproduction, and that their interpretations by the actors in any system will never be irrelevant to the history of reproduction.

I have argued that the difficulties experienced by culture theorists are to be expected, given that the aspects of culture which give it a group flavor—its temporal inertia and its spatial patterning—are incidental outcomes of the collective effects of generations of individual humans pursuing their personal reproductive interests, individuals who have nevertheless been under the constraints of continual social scrutiny from others always with somewhat different interests.

I have not suggested that culture precisely tracks the interests of the genes—obviously this is not true—but that, in historical terms, it does so much more closely than we might have imagined, and that for the future, our enlightenment on this point is almost certain to reduce the extent to which culture follows the interests of the genes and increase the extent to which it tracks, instead, our phenotypic interests as individual and social collectives. At least for humans unaware of their reproductive history, then, culture will remain paradoxically both the hand-

maiden of the genes and the obligate environment of their reproduction. In who understand their history, repr bypassed deliberately, and often with likely to see the substitution of a comb duction and increased attention to satisfactions. I see no possibility that th tion will inevitably engender against su remotely approach in pace the accele changes abetting them.

How does one begin to test this set of isms reproduce solely by producing and a their genes. The carriers of their genes are and nondescendant relatives. Organisms evolved to be altruists, whose beneficence, ronments of history, is eventually directed Indeed, our hypothesis has been that they maximally effective nepotists. Whatever acti gage in, then, it must be the first priority for t these activities to discover if possible how the ductive success in their current environments rive from activities that did so in past environm ment, or assertion, plainly assumes that evolu selection is universal, inevitable, and inescapa traits of organisms owe their existence and the process. Nowhere among living creatures, I su doubt be cast on this premise except from igno from absence of data rather than from data. In c seems to me that, in the 120 years since Darwin, proof has shifted unequivocally to those who v any other argument. (Relationships between evolu ory and the understanding of human behavior are sued by an increasing number of investigators, w degrees of success. The reader who wishes to compa ferent approaches, topics, and opinions will profit fror ing the following publications in addition to those where in this book: Barash, 1977; Barkow, 1978; Capl expect that patterns of kinship will be centrally important in human societies, and that humans will behave in predictable ways toward different genetic kin. We expect keen ability to sense differences in the needs of kin, and to respond to them. Similarly, we may expect cleverness in detecting availability of nepotistic benefits, and in diverting them to personal benefit, even sometimes by deception. Such deception may take two forms: (1) efforts to insinuate one's self into the role of a closer than actual relative, and (2) efforts to exaggerate one's need or ability to use sought-after benefits in reproduction. In turn, we should become very sensitive to all such efforts at deception, and very clever at avoiding nepotism unlikely to yield maximal reproductive returns. We should expect that assistance to kin, on one's own terms, should represent one of the most potent of all sources of pleasure and satisfaction.

It seems to me that passages like the three that follow—taken from the writings of anthropologists unaware of the arguments presented here—represent powerful support for the view that human sociality is derived through a history of differential genetic reproduction.

> In most primitive societies the social relations of individuals are very largely regulated on the basis of kinship. This is brought about by the formation of fixed and more or less definite patterns of behaviour for each of the recognised kinds of relationship. There is a special pattern of behaviour, for example, for a son towards his father, and another for a younger brother towards his elder brother. The particular patterns vary from one society to another; but there are certain fundamental principles or tendencies which appear in all societies, or in all those of a certain type. It is these general tendencies that it is the special task of social anthropology to discover and explain. [Radcliffe-Brown, 1924, p. 544]

> No worse affront can be hurled in the teeth of a Kurnai Australian than to call him an orphan; and the same is true of the Crow Indian in Montana. That so harmless a term should be resented as the most offensive imprecation seems strange, but

there is an explanation for it. Among the ruder peoples influence is often directly dependent on the greater or lesser number of faithful relatives. The kinless orphan is consequently damned to social impotence and considering aboriginal vanity it is natural that the vocabulary of vituperation should contain no more degrading epithet. It is therefore not only certain that neither the Kurnai borrowed from the Crow nor vice versa, but the reason for the observed parallel is clear from known facts of primitive life. [Lowie, 1920, pp. 11–12]

Around every person there is a circle or group of kindred of which such person is the centre, the *Ego,* from whom the degree of the relationship is reckoned, and to whom the relationship itself returns. Above him are his father and his mother and their ascendants, below him are his children and their descendants; while upon either side are his brothers and sisters and their descendants, and the brothers and sisters of his father and of his mother and their descendants, as well as a much greater number of collateral relatives descended from common ancestors still more remote. To him they are nearer in degree than other individuals of the nation at large. A formal arrangement of the more immediate blood kindred into lines of descent, with the adoption of some method to distinguish one relative from another, and to express the value of the relationship, would be one of the earliest acts of human intelligence. [Morgan, 1871, p. 10]

The diagram of relationships shown in figure 4 effectively conveys the fact that, normally, every individual lives out his life embedded in a network of near and distant kin. The importance of degrees of genetic relatedness is suggested by the number of different kinds of relatives with distinctive names. If we leave aside for the moment variations among societies, and consider only American society, the words on the diagram are communicative signals, well understood and widely used. All of these different names for different relatives are used within essentially every segment of society, even when large groups of relatives or clans are so powerfully cohesive that they behave like single families and use the word "family" or "clan" to denote the boundaries of kinship. Each individual is not only a member of a particular "family" or "clan" but in Ego's terms a particular kind of relative within the group.

It is especially relevant that individuals in human social systems, particularly those of close genetic relationship, are designated or named according to their particular relationship to Ego rather than solely in a fashion designating them as a member of a given group or society. Anthropologists argue that culture is a group phenomenon, and in some respects it certainly is; in modern society, designations of class, religion, race, or political party, in which individuals tend to lose their separate identities, are clear evidence of such group functioning. But the meticulous delineation of individual genetic and social relationships within human society also indicates quite clearly that the structure of human social systems, at least historically, has not been such as to erase individual functions completely or disallow competition at the individual level. The widespread familiarity of the information in figure 4 supports the validity of a biological approach to the analysis of human sociality emphasizing reproductive competition at the individual level. The reason is that every individual has a separate set of relatives, in terms of relatedness and in terms of needs, or ability to convert received benefits into genetic reproduction. The cultural anthropologist Anthony F. C. Wallace (1961) described as well as any what I would regard as an everyday consequence of this aspect of our genetic history:

> The humanist—the poet, the novelist, the dramatist, the historian—has tended to approach . . . with a sense of tragedy (or humor) . . . the paradox, so apparent to him, that despite the continuing existence of the culture and the group, the individual is always partly alone in his motivation, moving in a charmed circle of feelings and perceptions which he cannot completely share with any other human being. This awareness of the limits of human communication, of the impossibility, despite all the labor of God, Freud, and the Devil, of one man fully understanding another, of the loneliness of existence, is not confined to any cult of writers; it is a pan-human theme. [P. 130]

We all know that the word "family" usually refers to individuals indicated inside the first circle surrounding Ego—parents, offspring, and siblings (figure 4). All of these relatives, in a monogamous family, are related to Ego by the same amount, one-half, either exactly (parents and offspring) or on average (all others). Hence, there is a genetic correlate for the generalization "family," and for the tendency to refer to other individuals on the basis of whether or not they are members of the family. In general, throughout the human species, assistance is given and sex is avoided (except between the parents), unquestioningly, inside the family (Murdock, 1949).

Similarly, every individual recognizes that at some level toward the outside of the diagram he and his family tend to stop treating relatives in a special fashion, or even referring to them as relatives. Most people in a modern technological society, for example, may know of the existence, at least, of all of their first cousins, but few could count, let alone name, all of their second or third cousins, or their cousins once or twice removed (Schneider and Cottrell, 1975). At these points relationships are usually treated as trivial, although it is not irrelevant that the precise relationship of unusual, famous, or wealthy distant relatives is often remembered and repeated. In general, however, distant relatives, even of differing degrees, are combined under some label such as "distant cousins." In other words, variance in relatedness among associates increases beyond the limits of the nuclear (especially monogamous) family. Then, for practical purposes, the significance of distinguishing relatives decreases beyond some level, such as that of first cousins, because of low relatedness; these changes are signaled by tendencies to combine near and distant relatives under the general terms of "family" and "distant cousins," respectively. Between these two extremes relatives tend to be grouped according to (1) relatedness (i.e., sister or brother versus aunt or uncle versus cousin), (2) usual degree of dependency of relatives as compared to Ego (e.g., aunt or uncle versus niece or nephew), and (3) sex (e.g., aunt versus uncle, niece versus nephew). These kinds of groupings, which are not greatly different from those distinguished by early anthropologists such as A. L. Kroeber (1909), are obviously consistent with a Darwinian model (see also Murdock, 1949).

The first two points of evidence supporting an evolutionary interpretation of human social behavior, then, are that (1) *all societies operate as kinship systems,* and (2) *within these systems kin tend to be distinguished on the basis of degree of relatedness and relative dependency.* Both facts are entirely consistent with an inclusive-fitness-maximizing theory of sociality and culture, with selection effective principally at and below the individual level, and both are contrary to group-selectionist as well as nonevolutionary views.

Some Terminological Problems

Because the biological and social sciences have for so long gone their separate ways, it is important to note the differences between their concepts and approaches. Three major focal points of anthropologists and others analyzing human sociality are those commonly termed (1) kinship systems, (2) exchange systems, and (3) marriage systems, the last including the rules and patterns of incest avoidance. Social biologists may be led to assume that these systems are equivalent to their own interests in (1) nepotism, (2) reciprocity, and (3) the composite results of sexual competition, sexual selection, sex-differential parental investment patterns, and genetic outbreeding. While this assumption may have been correct at some times during anthropological history, or in the writings of some social scientists, it is generally an inaccurate interpretation of current investigations. The "kinship systems" of modern anthropologists are not merely systems of nepotism, but include both nepotism and reciprocity, as well as the results of sexual competition, sexual selection, parental investment patterns, and genetic outbreeding. Exchange systems and marriage systems are each usually discussed so as to include all of the same phenomena. This discordance is at the root of current disagreements over the applicability of evolutionary theory to understanding human history. When the argument is made that "kinship systems are not based on biology," it usually means that they are demonstrably not merely systems of direct nepotism. It does not follow, however, that human kinship systems do not accord with the principles of evolutionary biology, unless it could be demonstrated that systems of marriage and reciprocity are not developed so as to reflect differences of genetic interests as they have existed through history.

Disparities between the concepts of social and biological sciences may have come about partly because the sciences of anthropology, psychology, and sociology developed without the particular analytical approach that now seems appropriate to evolutionary biologists. Also contributory, I believe, are conscious and unconscious tendencies within human societies to mix nepotistic, reciprocal, and sexual interactions— for example, to employ terms and practices that ordinarily reflect genetic kinship to emphasize or guarantee reciprocity in situations explicitly not involving genetic kin. Here I am suggesting it is useful to interpret the systems examined by social scientists in terms of the systems conceived by biologists concerned with natural selection. Thus, each time I analyze some aspect of a kinship or other human system, I will try to dissect it into the components of sociality that seem important from a selectionist view of behavior. I have diagrammed the possible relationships among these different systems in figure 8.

Figure 8 implies that parental behavior is the primary evolutionary source (or preadaptive basis) of nepotism within the family, and that nepotism within the family is similarly the primary source of extrafamilial nepotism. Nepotism in turn is a likely source of systems of reciprocity (Alexander 1974, 1978c). Social cheating, or deception, which is potentially rampant in reciprocal systems, is also possible in nepotistic systems (especially by potential recipients of nepotistic benefits); hence, as with reciprocity in general, such deception may often originate in nepotistic systems.

relatives in figure 4. Especially when such designations are formal and institutionalized within societies, they are part of what anthropologists refer to as "classificatory" aspects of kinship nomenclature.

The widespread existence of such classificatory designations has been used to deny or deprecate the likelihood of a biological background for kinship systems. Some authors (e.g., Schneider, 1968, 1972; Sahlins, 1976a) have gone so far as to use the existence of classificatory designations to reject any notion that biological relatedness correlates significantly with kin terminology and patterns of social treatment.

I find the contentions of these authors mystifying, and I have taken an opposite stance (Alexander, 1977b, 1 'b). Suppose that I as an individual, or the members of my so generally, should wish to adjust the apparent social distar or the privileges and responsibilities, of some particular men bers or classes of members in the society. First, that such shifts are at all possible indicates a degree of regularity within society. Second, that terms commonly denoting classes of genetic relationship are employed to create the adjustments indicates that the regularity itself is based on genetic relationships. After all, if I should wish to draw inward, socially, a particular individual, what reason is there for me to use a kin term like brother except that brother ordinarily refers to someone with the close relationship that I wish to create? The existence and nature of at least some classificatory aspects of kinship thus support rather than negate the notion that genetic relationships and nepotism represent the basic cement of human sociality.

Other classificatory aspects of kinship also support this view. Thus, not every possible distinction is made. As already noted, relatives of the same degree but different dependency status, such as uncles and aunts versus nephews and nieces, *are* distinguished. Relatives derived through different routes, such as maternal and paternal relatives, are sometimes not distinguished, nor are the various kinds of cousins. Sometimes the sexes are not distinguished. Rather than automatically rejecting the possibility of functional explanations for such variations, however, we are challenged by their diversity to examine them for the possibility of explanations consistent with evolutionary theory. Not only have such biological analyses not been attempted, but satisfying nonbiological explanations of such cultural variations have yet to be devised. For example, it should be a matter of keen interest that kin terms for relatives that, according to Murdock (1949), are banned as sexual or marriage partners, either always (e.g., full siblings, parents, offspring) or nearly always (e.g., aunts or uncles, nieces or nephews, half siblings), distinguish those relatives by sex. Conversely, kin terms that apply to cousins, who are banned sexually much less often and sometimes are even prescribed marriage partners, usually do not distinguish the sexes.

To take another example, when the only appropriate marriage partner is a matrilateral cross-cousin and a certain individual has none, to designate some other individual may at first seem arbitrary and capricious. But this practice may keep the system of matrilateral cross-cousin marriages intact and thus serve the interests of those who benefit from its maintenance. It is not, then, necessarily biologically arbitrary or genetically irrelevant.

The nomenclature of kinship and sociality may also be employed at two or more levels. For example, in some societies an individual may call a distant relative or an unrelated person "brother." It does not follow, however, that the individual will necessarily treat the other like a true brother in every social circumstance, or, least of all, actually confuse the other with his true brothers. Conversely, it does not follow that true brothers will, frequently or ever, refer to each other by a term specifically designating siblings. Just as one may call anyone "brother," and otherwise treat him like a brother or not as circumstances dictate, one can also know his own brothers, and treat them accordingly, with sparse or no use of a specific kin term. The true significance of use of a kin term, or failure to use one, may thus be confusing to observers of a social group, and even to some participants in the system. Indeed, as everyone is aware, a term like "brother" may very well be used, quite deliberately, to deceive either the individual designated by it or someone else who may observe the designation.

Murdock (1949), in an excellent review of the use of kinship terms, points out that kin who are merely classified as being close are characteristically avoided in sex and marriage less often than real kin in the same categories. This fact by itself suggests that kin are typically given the classificatory designation of some closer relative, rather than of a more distant one. While a relative may be "disowned," the more common practice of classificatory designation has the opposite effect. The implication, it seems to me, is that (1) the symbols for degrees of relatedness are being used by individuals to adjust their social interactions with preferred (or reciprocating?) individuals, and (2) individuals have more control, or more interest, in shifting distant relatives into more favorable situations than in shifting disfavored relatives into less favorable situations.

Complications of these sorts certainly do not suggest that natural selection has not influenced the history of the structure of human sociality. Instead, their analysis reinforces the idea that such principles will finally put us on the correct route toward understanding our individual and collective tendencies and motivations.

It is almost impossible to find an anthropological description of a kinship system that contradicts the arguments made above, even though the likelihood of a match to Darwinian predictions by chance alone seems vanishingly small. The disagreements appear, not in *data*, but chiefly in the *interpretations* by social scientists, from varying theoretical perspectives, of what they have found. The refinements of evolutionary theory described in the last chapter suggest that we must return so far as possible to the ethnographic data and ignore interpretations and rejections based on early views of natural selection or anthropological theories, now known to be faulty.

*It is fortunate that the leaders of both
cultural and social anthropology have left
behind such a rich heritage of descriptive
ethnography, for their legacy of theory,
however admirable for its ingenuity and
provocative quality, includes virtually
nothing of solid value for the future science of man.*
—George Peter Murdock, 1972, p. 22

BIOLOGICAL PREDICTIONS ABOUT HUMAN SOCIALITY

The following list is a review of twenty predictions I have
previously published and a continuation of the list. While some
of these predictions are trivial, taken individually, and some
may be faulted as circular because our immersion in the human
system of sociality already tells us they are true, collectively
they are significant, and the longer the list the more important
it becomes. Over one hundred such predictions, in one form or
another, can be located in this book.

1. When the abilities of potential recipients of nepotistic
benefits to translate such benefits into reproduction are equal,
then closer relatives will be favored over more distant relatives.
Murdock (1949) makes several remarks that leave little doubt
with respect to this prediction. On page 14 he notes that "some
of the intimacy characteristic of relationships within the nuclear family tends to flow outward along the ramifying channels
of kinship ties. A man ordinarily feels closer, for example, to the
brothers of his father, of his mother, and of his wife than to
unrelated men in the tribe or the community. When he needs
assistance or services beyond what his family of orientation
[that in which he was born and reared] or his family of procreation [that into which he has married] can provide, he is more
likely to turn to his secondary, tertiary, or remoter relatives
than to persons who are not his kinsmen." On page 15 he adds,
"A rule of descent affiliates an individual at birth with a particular group of relatives with whom he is especially intimate and
from whom he can expect certain kinds of services that he

cannot demand of non-relatives, or even of other kinsmen." On page 43: "Consanguineal kin groups are of particular importance, for a person ordinarily feels closer to his own 'blood relatives' than to those who are related to him only through marriage." On page 57: "However much [members of a kindred] may disagree or quarrel, they are expected to support one another against criticism or affronts from outsiders."

2. When genetic relationships of potential recipients of nepotistic benefits to a potential giver of benefits are equal, then relatives with the greater ability to translate benefits into reproduction will be favored. (These first two predictions, which represent the major assumptions of kin selection, are a restatement of Hamilton's 1964 generalization, quoted earlier; numerous subsidiary and more specific hypotheses flow directly from them.)

3. Relatives of the same degree are likely to be distinguished, nomenclaturally and otherwise, only when their abilities to translate benefits into reproduction are consistently different. (Thus, uncles and aunts are usually distinguished from nephews and nieces.)

4. Cooperativeness and competitiveness between particular sets of relatives, such as full siblings, may vary across essentially the entire spectrum of possibilities, depending upon their opportunities and needs to use the same resources (e.g., parental care or mates) and the value to each of having a cooperative individual available. (Thus, the phrase "sibling rivalry" simultaneously connotes a high level of competitiveness and implies that it has an unusual or surprising aspect.)

5. Because some kinds of genetic relatives, such as monozygotic twins, have been produced and reared only rarely during human history, no special social behavior consistent with their genetic relationship would have had an opportunity to arise. Monozygotic twins, then, might be expected to behave like siblings of very similar ages, although such siblings were probably also rather rare during human history as a result of inhibition of ovulation during lactation, postpartum sexual taboos, and infanticide.

6. The boundary of effective nepotism may be clarified by drawing some relatives at the boundary inward, nomenclaturally and otherwise, and by pushing others outward (hence, as already noted, some of the apparent inconsistencies in kin nomenclature).

7. Labels that usually designate kin may be applied to non-relatives in the course of establishing, stabilizing, or guaranteeing social reciprocity (hence, classificatory aspects of kinship).

8. Older offspring are likely in many circumstances to be reproductively more valuable than younger offspring, and to serve their parents' interests better in the course of serving their own interests because of their typical age and dependency relationships to their siblings (leading to primogeniture). Younger dependent offspring, on the other hand, may be given full attention with fewer reservations because of the diminished likelihood of additional dependent young (hence, may be "spoiled").

9. In polygynous societies male offspring with predictably great likelihoods of success will be favored over all others (Trivers and Willard, 1973; Alexander, 1974; Dickemann, 1979).

10. Offspring that are abandoned or destroyed are likely to have been of low reproductive value, either because of their phenotypic attributes or because of the timing or the circumstances of their appearance (Alexander, 1974; Dickemann, 1979).

11. Nondescendant relatives may sometimes be better avenues of reproduction than descendant relatives (leading in some cases to adaptive celibacy).

12. Because parent-offspring sociality has been continuous in human history, interpretations of adoptions and quasi-nepotistic behavior toward other species, such as dogs and cats, must be interpreted in terms of the alternatives available to the involved parties; and proximate motivations, or physiological causes, can only be understood as molded by a long-term selective history. An individual evolved to show parental behavior and cheated of the opportunity does not violate evolutionary predictions by showing it to either adopted children or pets.

13. Potential recipients of nepotistic benefits are usually in a better position to gain by cheating, than are potential givers, in the following ways: (a) by deceiving about the closeness of their relationships to potential givers of benefits, and (b) by deceiving about the extent of their ability to translate benefits into reproduction.

14. Relatives by marriage are in a particularly favorable position to gain by cheating their affluent in-laws, since they do not gain directly by the distribution of any benefits to their spouse's relatives (hence, the prominence of in-law jokes).

15. Men are physically more powerful than women, hence more likely to have their way in conflicts of interest; rules of kinship and marriage may be expected to reflect that difference (see chapter 4 and below).

16. Interactions between young individuals (dependents, or individuals with high reproductive value) and their very old relatives (with low reproductive value) remain nepotistic for the old individuals but assume the form of social reciprocity for the young individuals, since as individuals age they are less and less able to turn nepotistic benefits into reproduction.

17. Lowered confidence of paternity causes a man's sister's offspring to assume increased importance as recipients of nepotistic benefits, and in extreme cases to exceed his own spouse's offspring in this regard (hence, the importance of the "mother's brother" to dependent children in some societies—Alexander, 1974, 1977b; Greene, 1978; Kurland, 1979; and see below).

18. Erratic confidence of paternity may lead to emphasis upon phenotypic attributes of putative offspring in determining whether or not to accept them as suitable objects of paternal care (hence, perhaps, the frequent attention to the question of whether or not a baby resembles its father).

19. Sexual activity with spouse's siblings or sibling's spouse leads to asymmetry in the genetic relatedness of putative cross- and parallel-cousins (hence, their separate treatment and nomenclature, concentrated in polygynous societies—Alexander, 1974, 1977; Greene, 1978; and see below).

20. Regarding in-laws, after appropriate devaluations for

likelihood of defection or use of resources for their relatives not shared with Ego, because of the descendants they share with Ego (or are expected to share), they may be expected to be treated like the relative for which they are named and to which they correspond (brother-in-law as brother, daughter-in-law as daughter, etc.). Analyses of social interactions ignoring the reproductive significance of marriage, and this logical extension of kin selection (e.g., Sahlins, 1976a) can only be misleading and wrong (I acknowledge the assistance of William G. Irons with regard to this point).

21. Inheritance is most likely to be evenly distributed between male and female offspring (i.e., to be bilateral) when marriages are monogamous or sororally polygynous and least likely to be so when marriages are nonsororally polygynous. This is so because women in polygynous marriages are likely to have to share resources with other wives, and if they are not sisters the resources inherited by daughters will not all be directed to the inheritance-transmitters' grandchildren as they would be if inherited only by sons (patrilineal inheritance). The distributions of patrilineal and bilateral inheritance are as these statements predict (Borgia and Alexander, ms.).

22. Parents should prefer to pass inheritance to daughters when daughters can control the resources since low confidence of paternity (compared to confidence of maternity) makes a daughter's children closer relatives on average than a son's children. This may help explain correlations of the avunculate, matrilateral cross-cousin marriages, and matrilineal inheritance, and it is also supported by the greater prevalence of matrilateral as opposed to patrilateral cousin marriages: a man can pass inheritance more profitably to children through his daughter and his nephew than to children through his son and his niece (see pp. 183–91).

23. Individuals should rather pass inheritance to offspring married to relatives than to offspring married to nonrelatives. Thus, when cousins are, for whatever reasons, usual marriage partners, each parent will prefer that children marry its own rather than its spouse's relatives. Since men are likely to win,

parallel cousin marriages are on this basis more likely through brothers than through sisters. This general pattern will recur whenever known relatives, whether cousins or not, are the usual marriage partners, and when heritable goods are such (i.e., nonpartible) as to make it more profitable to be closely related to a few descendants than to be distantly related to many.

24. Larger social groups provide more opportunities for social cheating, and on this basis are likely to be typified by greater emphasis on reciprocal transactions, as opposed to nepotism (Alexander, 1978c, and see below).

25. Other selective contingencies, such as the value to individuals of their own social group's (eventually nation's) not losing out to its competitors, also may cause reciprocity rather than nepotism to become the cement of group-living (hence, perhaps, a drop in the importance of extrafamilial nepotism, and the removal of the upper limits of group size imposed by nepotism; and perhaps also the particular forms, intensities, and influences of religions in successful large nations—Alexander, 1978d, and see below).

Earlier I implied that most anthropological discussions are so interpretative that their dependence on group-level or nonreproductive functions makes them difficult to use in discussions based on a theory of inclusive-fitness-maximizing by individuals. The single outstanding exception is Murdock's (1949) analysis of data from the Human Relations Area File. Although he later rejected the group-benefit approach to cultural analysis (see Murdock, 1972), in 1949 Murdock developed his discussions around such a theory, continually speaking of what the "society" must do for its own good. He even began his book with the implication that the nuclear family was established by society for the purpose of its own perpetuation, and that the production of children is an operation of the nuclear family designed to benefit the society as a whole. Nevertheless, Murdock's book is filled with descriptive information and data on the biology of the human species that can be used to advantage by anyone with a different theoretical approach. Any data or conclusions of Murdock's that support an interpretation based on natural selection of genetic replicators were not derived because of this approach. Yet his book is filled with conclusions that could have been predictions from an evolutionary approach. I have here listed some of these conclusions, which I believe strongly support my approach; then I have discussed some of his conclusions that appear at first to be inconsistent with an evolutionary interpretation.

Some Statements from Murdock Consistent with Evolutionary Theory

1. "The nuclear family is a universal human social grouping . . . it exists as a distinct and strongly functional group in every known society . . . the one fact stands out beyond all others that everywhere the husband, wife, and immature children constitute a unit apart from the remainder of the community" (p. 2). This finding accords with the fact that parenthood is the sole or chief mode of reproduction in all nonhuman species. Although later investigators call into question some of Murdock's criteria of the nuclear family (e.g., Laslett and Wall, 1972; Schneider and Smith, 1973), their restrictions seem to me to validate rather than weaken his conclusions.

2. "Marriage exists only when the economic and the sexual are united into one relationship, and this combination occurs only in marriage. Marriage, thus defined, is found in every known human society. In all of them, moreover, it involves residential cohabitation, and in all of them it forms the basis of the nuclear family" (p. 8). Only the "residential cohabitation" part of this statement seems questionable (e.g., Alland, 1972).

3. "The primary effect of a rule of residence is to assemble in one locality a particular aggregation of kinsmen with their families of procreation" (p. 17). "In the great majority . . . 140 out of the 187 in our sample . . . nuclear families are aggregated . . . into . . . clusters of two, three, or more . . . which commonly reside together and maintain closer ties with one another than with other families in the community. . . . Socially, the nuclear families thus associated are almost invariably linked to one

another not only by the bond of common residence but also through close kinship ties" (p. 23). These facts are obviously consistent with evolutionary views of nepotism.

4. "The data from our 250 societies . . . reveal that some form of consideration ordinarily accompanies marriage when residence rules remove the bride from her home. The payment of a bride-price and the giving of a sister or other female relative in exchange for a wife are almost exclusively associated with patrilocal residence. Bride-service [assistance to her family] normally accompanies matri-patrilocal residence and is also common under matrilocal residence" (p. 20). A man pays to remove a woman from the vicinity of her relatives and he pays with direct assistance when he moves in with her family. Whatever he pays may well be viewed as a guarantee of his continued attention to her: if she returns or is abandoned he loses the bride-price or service; if the marriage continues he may well expect it back in direct inheritance or assistance. Even if this specific interpretation applies only to some cases, the general point seems obvious.

5. "193 societies . . . are characterized by [permitting] polygyny, and only 43 by monogamy and 2 by polyandry" (p. 28). This distribution is what one might expect for a mammal that lives in groups with both the male and female being highly parental—although humans are the only such mammal species.

6. "Most of these . . . instances [of plural unions in otherwise monogamous tribes] [are cases in which] . . . a special exception is made for men whose first wives prove barren" (p. 27).

7. "Group marriage . . . appears never to exist as a cultural norm" (p. 24). Nor does it in nonhuman species, although females are promiscuous in many species with little or no male parental care.

8. "The polyandrous family occurs so rarely that it may be regarded as an ethnological curiosity" (p. 25). It is similarly rare in all mammals, and even in birds.

9. "Non-fraternal polyandry is exemplified by the Marquesans, among whom a number of unrelated men join the household of a woman of *high status* and participate jointly in eco-

nomic responsibilities and sexual privileges" (p. 26; emphasis added). It would seem that this situation could prevail only in a stratified society, and it parallels those circumstances in birds in which females that defend superior territories acquire harems of males to incubate their eggs (Jenni, 1974).

10. "Co-husbands among the Todas occupy one house when they are brothers, but in the occasional instances of non-fraternal polyandry they maintain separate dwellings where they are visited in rotation by their common wife" (p. 26). See statement 12 for the parallel situation in co-wives.

11. "Our data show preferential sororal polygyny [as opposed to nonsororal] to be exceedingly widespread" (p. 31). One expects sisters to cooperate more successfully, and sororal polygyny allows parents to pass resources to their daughters without their being redistributed so as to pass to someone else's grandchildren.

12. "Among the Crow, Sinkaietk, and Soga . . . co-wives regularly reside in the same dwelling if they are sisters. . . . In 18 of the 21 societies with exclusive sororal polygyny . . . co-wives live together. . . . In the majority of societies with nonsororal polygyny (28 out of 55) co-wives occupy separate dwellings, and in many of the cases where they share the same house it is a large dwelling in which they are provided with separate apartments" (p. 31).

13. "Generalized obligatory [sexual] regulations such as wife lending and sexual hospitality are . . . exceedingly infrequent" (p. 264).

14. "Special social status may . . . be associated with permissive [sexual] regulations . . . [as] the right of a feudal lord, priest, or other male in a position of authority to have sexual intercourse with a bride on her wedding night before her husband is allowed access to her. Another instance is the special prerogative of violating the usual incest taboos which certain societies accord to persons of exceptionally high status" (p. 266).

15. "In the case of [sexual avoidance of] MoBrWi [mother's brother's wife] . . . at least four and perhaps six of the seven instances of avoidance and three of respect occur in conjunction

with cross-cousin marriage with MoBrDa [mother's brother's daughter]. . . . Moreover, in three of the five instances of joking the maternal uncle's wife is a potential levirate spouse, like BrWi [brother's wife], and is accorded similar treatment" (p. 281).

16. "With the exception of married parents, incest taboos apply universally to all persons of opposite sex within the nuclear family" (p. 284).

17. "Incest taboos do not apply universally to any relative of opposite sex outside of the nuclear family" (p. 285).

18. "Incest taboos and exogamous restrictions, as compared with other sexual prohibitions, are characterized by a peculiar intensity and emotional quality" (p. 288). I suggest that this is most true when very close relatives are involved (see statement 19), which means that all involved parties are likely to benefit from incest avoidance so that no deception is involved (see also p. 195).

19. "Incest taboos tend to apply with diminished intensity to kinsmen outside of the nuclear family" (p. 286).

20. "Where . . . property in agricultural land, large permanent houses, localized fishing or grazing sites, and other immovables constitutes a major form of wealth, especially if the most efficient unit for economic cooperation is larger than the nuclear family, each new family of procreation will tend to cleave to and assist the family of orientation of the spouse who can expect to share in the inheritance of such property. In this way, probably, the prevailing forms of property and the mode of its inheritance can predispose a society to a particular form of extended family" (p. 37).

SOME STATEMENTS FROM MURDOCK
THAT DO NOT SUPPORT EVOLUTIONARY THEORY

To be consistent I must note here that a few statements from Murdock are apparently inconsistent with evolutionary theory. I have located six. For each I have either offered a possible explanation consistent with evolutionary theory, or described what seem to me the appropriate questions to ask; otherwise these questions are untested.

1. "In Africa and elsewhere . . . it is common for the illegitimate children of a married woman by another man to be unquestioningly affiliated by patrilineal descent with her husband, their 'sociological father' " (p. 15). One wishes to ask (a) whether such children are valuable to the sociological father, either as labor or as a source of bride-price, and (b) what other benefits may accrue to the man who accepts such children. Does he thereby retain a wife of value that he would otherwise lose? Does he maintain ties with an important group of affinal allies? Is the real father commonly a close relative such as a brother? Are such events reciprocal? How is the child actually treated by him? The cases to which Murdock refers are probably those in which a girl's mother selects for her a lover who fathers her first child. In such cases we need to know the status of the lover, the eventual distribution of inheritance, the extent of knowledge of paternity in all cases, the behavior of the real father toward the first child, and more. Without such kinds of information there is no way to analyze the apparent paradox.

2. Murdock says that the three primary rules of descent are employed so as to *restrict* the kin group of a child so the kin group will not "become coextensive with the community, or even the tribe, and . . . lose its significance" (p. 44). If this were the reason why *close relatives* restrict the kin groups of a child, then the predictions of correlates of different descent systems would surely be at variance with those from evolutionary theory. From the point of view of the child and its close relatives, significance of the kin group would only be lost if variations in relatedness were lost or distorted; from the viewpoint of others, though, especially nonrelatives, it would be disadvantageous if any individual successfully claimed kinship with all members of the group, while others could not. Of course something like this happens with chiefs, or other persons of great influence, because everyone believes he can gain by claiming relationship to such persons.

3. "Most societies impose a taboo against sexual intercourse

during a woman's menstrual periods, during at least the later months of pregnancy, and for a period immediately following childbirth" (p. 266). Sexual taboos following childbirth delay the next pregnancy and correlate with physiological inhibitions of ovulation during nursing. As such they seem to be mechanisms of greater success in reproduction by the spacing of childbirth. Such taboos during late pregnancy have the obvious possible correlate of reducing the likelihood of interference with the pregnancy: similar advice is given to couples by physicians in modern societies, and especially to couples in which there is a high likelihood of "spontaneous" abortion. Taboos during menstruation, on the other hand, are widespread and represent a difficult topic which deserves more discussion than I can give it here. It is even possible that the copious menstrual flow of women has evolved to be a signal of nonpregnancy. Nor do we understand the tendency of women to synchronize their menstrual periods with other women with whom they interact closely (McClintock, 1971). These problems must be taken into account when trying to assess the significance of different rules and attributes associated with menstration in different societies.

4. "The Azande and the Kiwai enjoin copulation during pregnancy in order to promote the development of the fetus" (p. 267). This behavior seems precisely opposite to that described above, therefore represents an excellent topic for further analysis. All that can be said here is an evolutionary approach predicts that, upon analysis, these two behaviors will prove *not* to be exact opposites of one another. One wishes to answer questions like: from whom does the impetus to copulate during pregnancy come, and what may be its significance, in different cases, in maintaining the bond between the father and mother?

5. "The commonest restrictive [sexual] regulation . . . is the requirement of celibacy and often also of chastity in priests and other religious functionaries" (p. 266). I have already discussed a possible evolutionary explanation.

6. ". . . incest taboos, in their application to persons outside of the nuclear family, fail strikingly to coincide with nearness

of actual biological relationship" (p. 286). The evidence is over-whelming that genetic outbreeding is beneficial to those who practice it. Sexual avoidance of closest relatives is the basis for the concept of incest avoidance. The correlative failure Murdock cites characterizes polygynous societies in which marriages are arranged, and are not necessarily in the best interests of the marriage partners. Not only are genetic relationships less certainly ascertained in polygynous societies—partly because of low confidence of paternity—but using the concept of incest to assist in arranging marriages may sometimes cause those being manipulated to believe that their interests are being served through outbreeding, or at least through avoiding behaviors that others have presented as abhorrent and unspeakable. Arrangement of marriages presumably serves the interests of those doing the arranging, sometimes by diminishing competition among relatives for the same mates, and sometimes by solving the crucial problem of finding mates for one's young male relatives in a polygynous society by involving the maintenance of alliances.

ANALYSIS OF TWO OTHER PHENOMENA
APPARENTLY CONTRARY TO EVOLUTIONARY THEORY

Several years ago I searched the anthropological literature for the most provocative and outstanding apparent contradictions of an evolutionary view of human behavior. I located two: the phenomenon of the avunculate, or "mother's brother," and the asymmetrical treatment of cousins. In 1974 I suggested tentative hypotheses, and in 1977 I continued the analyses from these preliminary hypotheses.

In these earlier papers I was examining human social behavior and cultural rules and practices as a rank outsider. I was stepping into the analyses in a piecemeal fashion partly because I had no broad knowledge of culture and partly because there was yet no general subtheory, under evolutionary theory, explicitly applicable to human problems. No such theory yet exists, but I have explored further with regard to cousin treatment and the avunculate. This further exploration has not negated any of the 1974 and 1977 conclusions, but it has convinced me that my earlier analyses include only a small portion of the problems of explaining cousin treatment and the avunculate, and it has also suggested to me a manner of approach to the problem of general theories about cultural patterns. Here I shall first present modified versions of my earlier explanations of the avunculate and cousin treatment, then add to them a more detailed discussion of cousin marriages, and finally explain how I think this approach leads toward a useful general theory.

Mother's Brother

In many societies, paternal responsibilities are assumed and benefits dispensed, not so much by one's putative father or mother's spouse, as by a particular uncle, the mother's brother; or, in some societies, mother's brother is at least an important dispenser of such benefits. This relationship, usually termed the "avunculate," is often a general feature of a society. It has always puzzled students of culture, although various explanations have been offered (e.g., Radcliffe-Brown, 1924; Harris, 1971). A close examination of the genetic relationships involved reveals the following curious fact, of significance for a Darwinian approach, and illustrated in figure 9 (see also Alexander, 1974, 1977b; Greene, 1978; Kurland, 1979): *lowered confidence of paternity leads to asymmetry in one's genetic relatedness to putative nephews and nieces.*

The offspring of a man's full sister are one-quarter like him in genes identical by descent; those of his half sister are one-eighth like him. Only by a remote accident of meiosis or an almost equally remote mistake in maternity can the offspring of man's sister be totally unlike him (the latter mistake is not quite so remote in technological societies in which women give birth in large hospitals while under anesthesia). His spouse's offspring, on the other hand, are either one-half like him (in genes identical by descent) or totally unlike him; they will be intermediately related to him only

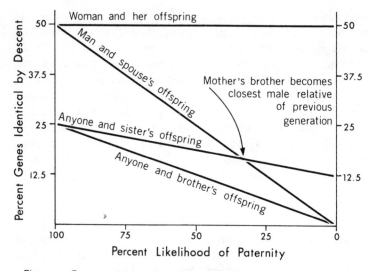

Figure 9. Genetic relationships with putative offspring, and with various kinds of nephews and nieces, with variations in confidence of paternity. Lowered confidence of paternity may be caused by such circumstances as husbands and wives living apart or long absences of husbands on hunting or military trips. Low confidence of paternity may be expected to lead to reduced paternal care by mother's spouse, thus increasing the value of care given by mother's brother. Tendencies by men to discriminate among their spouse's offspring on the basis of resemblance to themselves, or on the basis of other evidence of philandering by the spouse, will lead to the favoring of some of spouse's offspring over others, and to an increased favoring of sister's offspring over those rejected. Hence, the point at which mother's brother becomes prominent in a society can actually be moved far to the left of the intersection of the lines indicating relationships to spouse's and sister's offspring. The location of this point will also be influenced by the value of paternal care, as opposed to maternal care, and by differentials in male ability to dispense paternal benefits (e.g., if rich uncles are likely, tendencies to seek their assistance might also be likely).

To simplify the diagram I have kept confidence of paternity and confidence of full (versus half) sibship equal; they need not be.

if his spouse has philandered with one of his relatives. As a man's confidence of paternity diminishes, therefore, his sister's offspring become more important to his reproduction compared to his spouse's offspring. Similarly, the brother of a woman whose husband has a low confidence of paternity becomes a more appropriate source of parental benefits to her offspring, who may otherwise suffer from lack of paternal assistance.

If low confidence of paternity is general throughout a society, a man's sister's offspring, because of the high confidence of maternity, can become his closest known relatives in the next generation. Thus, if paternity is on average correctly ascertained only one-fourth of the time, then a man's spouse's offspring will average one-eighth like him (and three of four will be totally unlike him), while his sister's offspring will average five thirty-seconds like him, and all will approach this degree of overlap. In making these calculations one has to take into account that lowered confidence of paternity will also lower the average relationships between brothers and sisters. I have assumed that if likelihood of paternity is one-fourth, then likelihood of siblings having the same father is one-fourth; other assumptions are also possible.

The effects of society-wide lowering of confidence of paternity are different for a man's brother's offspring because his brother will suffer equally from lowered confidence of paternity; and for a woman's siblings' offspring since women will retain high confidence of maternity and will always be more closely related to their own offspring.

To summarize, if a man lacks confidence of paternity, his nieces and nephews may be his closest relatives in the next generation. This alone does not mean that such relatives are the most appropriate targets of his parental care, since if his low confidence is unusual (i.e., other males are confident of their paternity), his nieces and nephews may be expected to have fathers willing to care for them. I repeat here that none of these calculations need be conscious to individuals who nevertheless behave as though they are.

Whenever general living conditions or other society-wide circumstances lead to a general lowering of confidence of paternity, only a man's sister's offspring, among all possible nephews and nieces, can become his closest relatives in the next generation. Moreover, children in such societies will generally fail to receive intense parental care from their mother's spouse. In consequence, so long as adult brothers and sisters tend to remain in sufficient social proximity that men are capable of assisting their sisters' offspring, we can predict that *a general society-wide lowering of confidence of paternity will lead to a society-wide prominence, or institutionalization, of mother's brother as an appropriate male dispenser of parental benefits.* We can add that to the extent that paternal care is more appropriately directed to males, mother's brother may be expected to attend more to sister's sons than to sister's daughters.

For an example of living conditions leading to lowered confidence of paternity, associated with a prominence of mother's brother, consider Alland's (1972) account of the Abron in Africa (it is worth noting that none of the authors cited here in any way supported—or was even aware of—the argument I am making):

> The married couple stays apart. The man is in residence with his father (or in some cases with his maternal uncle). . . . The woman lives in the house of her mother and her mother's sisters, with her siblings and the children of her mother's sisters. . . . If a man has two or more wives, they will live in different houses, since a man is not permitted to marry sisters. . . . On the death of the last surviving senior male of the same kin group related through a line of women . . . [the ownership of his house] passes on to the eldest son of the eldest sister of these men. This follows the standard inheritance pattern of the Abron which, as I have already indicated, is matrilineal (from man to man through a line of related women). The pattern of inheritance extends to land holding as well. A man does not inherit from his father but from his maternal uncle. . . . In many such societies the maternal uncle is occupied with the discipline and the father is free to play a supportive role. . . . [P. 105]

A parallel relationship between prominence of mother's brother and confidence of paternity is clear in the writings of Fortune (1963) on the Dobu of New Guinea:

Each villager, male or female, owns a house site and a house. . . . The husband in every marriage must come from another village than that of his wife. His house site is in one village, his wife's house site is in another. . . . [A] father bequeaths his house site to his own sister's son. His own son inherits house site and village status from his mother's brother in his mother's village [p. 2]. . . . The couple with their children live alternately in the woman's house in the village of the woman's matrilineal kin, and in the man's house in the village of the man's matrilineal kin [pp. 4–5]. . . . "Those-resulting from marriage," if they are men, are always abnormally uneasy about their wives' fidelity. Now when a woman is in her own village, she has her kin next door and only too ready to eject her husband if he dares to lift a hand against her, or use foul language to her. She has no great dependence on her husband for care of her children, since a woman can nearly always get a new husband for future help, and her brother ultimately provides for them in any case. Consequently she behaves very much as she likes in secret [p. 6]. . . . Dobuan folklore is full of husbands pathetically packing up their goods and going home to their mothers and sisters after a child has informed them that their wife has been consorting secretly with a male member of a distantly related *susu* of her own village [p. 7]. . . . Jealousy normally runs so high in Dobu that a man watches his wife closely, carefully timing her absences when she goes to the bush for natural functions. And when it is the time for women's work in the gardens here and there one sees a man with nothing to do but stand sentinel all day and play with the children if any want to play with him [p. 7]. . . . The chances of divorce are high. The oldest Dobuan in my main genealogy had had eight successive marriages, one of the youngest men in the genealogy had had four, one other youth, three, and this is fairly typical of an overwhelming majority of Dobuans [p. 9]. . . . A Dobuan tends to think of his sister first of all his ties when the question of breaking ties arises. Dobu practices the avunculate, inheritance from mother's brother to sister's son, but this is from no great sentiment between a man and his sister's son. Rather the sister enlists the brother in the interests of her children [p. 62]. . . . Every woman claims by right the inheritance of her brother for her male children [p. 3].

Similarly, the correlation of lowered confidence of paternity and the avunculate is prominent in some of the matrilineal societies discussed by Schneider and Gough (1974), and in some other cases of prominence of mother's brother (Murdock, 1967).

Fortune's description of the Dobu raises the question of how the institution of mother's brother became prominent in different societies, and how it acquired its relationships in some cases to matrilineality and lowered confidence of paternity. Any woman might be expected to turn to her brother for assistance upon loss of child support by a spouse as a result of separation or divorce. She seems less likely to obtain adequate assistance from a brother who is encumbered with full care of his own offspring. This suggests that shifts from paternal care to the avunculate where they have occurred could have involved (1) young and childless brothers, (2) brothers estranged from other wives, (3) unusually wealthy or powerful brothers, or (4) brothers who were for other reasons unlikely to be successful with their own offspring (e.g., see Dickemann, 1979), contributing to the care of their sister's offspring. Such a practice would in certain cases (e.g., the Dobu) relieve women of the necessity of maintaining less than optimal marriages, which in turn could reinforce the strength of the avunculate and the solidarity of kin relationships through mothers, and further weaken the institution of marriage. It is difficult to see how a woman could gain from her spouse's having a low confidence of paternity as such but sometimes she might benefit from a lessening of constraints imposed upon her behavior. In any case, she would not be as constrained to provide confidence of paternity if she could depend upon her brothers and other relatives for child support and other assistance.

Men would ordinarily gain reproductively from being able to tend their own offspring rather than those of sisters. But under certain conditions—as when living conditions, absence from the household as a result of military requirements, or other factors cause lowered confidence of paternity—it is easy to understand how shifts toward prominence of mother's brother, matrilineality, fragility of marriage bonds, and lowered confidence of paternity might go together, in patterns entirely consistent with a Darwinian model of human sociality. I emphasize that men may sometimes gain from favoring juveniles less closely related to them than their own offspring, and that much nepotism may be dispensed to sisters' offspring in ordinary situations (owing to the generally greater potential for power among males), causing brothers generally to assume responsibility for sister's welfare and that of her family. These circumstances commonly may occur without the extreme shift in relatedness that I have emphasized, a shift in which sister's offspring are actually favored over spouse's offspring. Prominence of the avunculate obviously does not require that a man's sister's offspring average a closer relationship to him than his spouse's offspring.

The avunculate is more widespread than is dramatically lowered confidence of paternity. Even if quite low confidence of paternity correlates closely with a prominent avunculate, it seems clear that not all behaviors classified under the avunculate have the same explanation. Murdock (1949, p. 35) points out that in the case of the avunculocal extended family, even prepubertal boys may leave their parental homes, going to live with a maternal uncle in another village. If, in such cases, these boys marry their uncle's daughter (MoBrDa), this system amounts to prepubertal adoption (and supervision and guidance) of a daughter's husband, on the one hand, and assurance of a wife for a still prepubertal son on the other. We may suspect, from this, that such systems are most likely under polygyny (because of enhanced sexual competition among males) and in societies with particular kinds of heritable goods that make profitable cross-cousin marriages facilitated by avunculocal residence. Confidence of paternity is not necessarily involved in this particular form of avunculate, but asymmetries of power and genetic relationships are, as discussed following the next section. Flinn (in press) has also shown that, independent of paternity uncertainty, frequency of divorce and remarriage influences a man's parental behavior.

Asymmetrical Treatment of Cousins

Cousins are the offspring of siblings. Offspring of full siblings, which share both parents, share on average one-eighth of their genes, identical by descent; offspring of half-siblings share one-sixteenth of their genes (figure 4).

Cousins are sometimes divided into two kinds, offspring of siblings of the same sex, called *parallel cousins,* and offspring of siblings of different sexes, called *cross-cousins* (figure 10). In technological societies derived from Europe (such as our own),

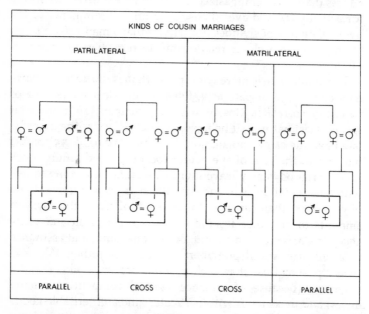

Figure 10. The four kinds of cousin marriages. In human societies in general, cross-cousin marriages occur more frequently than parallel cousin marriages, parallel cousin marriages more often through brothers than through sisters, and matrilateral cross-cousin marriages more often than patrilateral cross-cousin marriages. Cousin marriages tend to be the rule when marriages are rigidly arranged and subgroups in the form of clans or lineage groups are emphasized. When such subgroups are de-emphasized, arrangement of marriages diminishes and cousin marriages occur less frequently and may be disfavored or even forbidden.

cousins are usually not subdivided in this fashion, causing these terms to be novel for most persons other than anthropologists; but cousins are so divided in many other societies and in many different parts of the world (Murdock, 1967).

In general, close relatives are more likely to be objects of nepotism and less likely to be objects of sexual behavior than are more distant relatives. In the case of cousins there is a general correlation between the distinguishing of parallel and cross-cousins, the assigning of asymmetrical nomenclature, and their differential treatment, at least in regard to marriage. In many societies which distinguish parallel and cross-cousins, parallel cousins are termed siblings rather than cousins, and marriages between them, but not between cross-cousins, are either forbidden or discouraged. Although nepotistic behavior with regard to cousins has not been as prominently analyzed as has marriage behavior, the nomenclatural asymmetry implies that parallel cousins also favor one another in nepotism over cross-cousins.

As is typical of human social behavior, these practices are not always the rule. Thus, the fact of asymmetry of treatment and nomenclature, and the existence of actual reversals in a few societies (with parallel cousins rather than cross-cousins favored in marriage), figure prominently in arguments about the inappropriateness of assuming a biological background for kinship nomenclature and behavior. Since parallel and cross-cousins average the same degree of genetic similarity, it is asked why they should be treated differently (e.g., see Lowie, 1920; Levi-Strauss, 1969).

This question provides a good example of how the biological approach can be useful in understanding human cultural practices. First, we may ask whether the assessment of genetic overlap is really correct for the societies in question. It will be correct, of course, only when the true (genetic) parents are the same as the functional (social) parents. They are not always the same, however, and in a given society some kinds of deviations may be more likely than others. In a monogamous society with carefully kept records, for example, deviations may be few, and have little effect upon cousin marriages. So in societies with long-standing monogamy we expect parallel and cross-cousins to be treated the same.

But many societies, even today—and nearly all societies without advanced technology—permit polygynous marriages, or did so within recorded history. The antiquity of group living in humans, the universality of polygyny (sometimes deriving from promiscuity) in multi-male primate social groups, and the prevalence of polygyny even recently (Murdock, 1949, 1967) all indicate that polygyny has likely been a prevalent marriage system during most of human history. Moreover, in most polygynous societies, the prevalent form is sororal polygyny—that is, the different wives of one man may be sisters to one another. Additionally, as a consequence of polygyny, many less powerful and younger men are mateless. In this situation, men who are older and more powerful may accumulate more wives than they can actually retain and may pass them to younger brothers still unable to acquire wives on their own; this behavior has been reported in the Yanomanö Indians of South America (Chagnon, 1968), and in Australian Aboriginals (Money et al., 1970).

In a sororally polygynous society, the offspring of sisters will often be half-siblings. If brothers often share wives, or if older brothers transfer wives to younger brothers, the offspring of brothers will also average a closer relationship than will cross-cousins, more uncertainty will exist as to who has fathered particular babies than as to who has mothered them, and the potential errors will involve brothers.

Murdock (1949) makes three points relevant to this argument. First, "in a majority of the societies in our sample for which information is available a married man may legitimately carry on an affair with one or more of his female relatives, including a sister-in-law in 41 instances" (p. 6), and second, "the *levirate* is a cultural rule prescribing that a widow marry by preference the brother of her deceased husband. . . . The *sororate*, conversely, is a rule favoring the marriage of a widower with the sister of his deceased wife. . . . Both the levirate and the sororate are exceedingly widespread phenomena" (p. 29). Finally, "the most illuminating of privileged [sexual] relationships . . . are those between siblings-in-law of opposite sex. . . . Nearly two thirds of the sample societies for which data are available permit sexual intercourse after marriage with a brother-in-law or a sister-in-law" (p. 268).

In polygynous societies, then, people who seem to be parallel cousins may actually have the same father, and in fact they "are usually themselves called siblings in primitive languages" (Lowie, 1920, p. 26). Parallel cousins in such societies will thus on the average be genetically more closely related than cross-cousins (Harris, 1971; Alexander, 1974, 1977b; Greene, 1978). This fact must somehow be explained away before we can dismiss the differences in incest rules for cross- and parallel-cousin unions as genetically meaningless or owing solely to the particular whims or idiosyncrasies of different cultures.

Numerous asymmetries in genetic overlap between paternal and maternal relatives are possible, and they may correlate closely with tendencies to distinguish them in ways that European monogamous (hence, symmetrical) societies do not. It is worth stressing that the actual origins of such rules in any society, or the particular (proximate) reasons given for their existence, are not in themselves sufficient to cast doubt on the significance of such correlations. They may be as irrelevant to the question of reproductive significance or function as is, say, variation in the ontogenetic background of two bird songs to the fact that each song protects the territory of its possessor.

If incest avoidance is a reason why cross-cousin marriages are favored over parallel cousin marriages, then we can make several predictions about symmetry and asymmetry in the encouragement and discouragement of cross-cousin versus parallel cousin marriages, and likewise about marriages between parallel cousins with brother-fathers and with sister-mothers.

In monogamous societies in which marriages usually last a long time, or for the lifetimes of the partners, parallel and cross-cousin marriages should be most symmetrical. For at least two reasons this may not be so: (1) monogamy may be recent and (2) communal living of an asymmetrical sort (e.g., brothers with their wives in the same household, or sisters with their husbands) may lead to a greater likelihood of parallel cousins being half-siblings. The reverse can only be true (putative cross-cousins being half-siblings) if brother-sister incest occurs. If the argument is valid that culture is reproductively adaptive, then only when incest rules do not promote outbreeding should cross-cousins be called siblings when parallel cousins are not.

In cases of ancient monogamy, one should not expect parallel and cross-cousins to be distinguished at all. This is true of the Andaman Islanders (Radcliffe-Brown, 1922) and all of Murdock's (1967) twenty-eight societies in or derived from modern Europe.

A widespread failure of symmetry in cousin treatment in the expected direction (whether in marriage, nomenclature, or benefit-giving) might, on account of a human history of polygyny, alone support the notion that reproductive history is important in interpreting human culture. Asymmetry should be most common in systems of sororal polygyny in which men are permitted to make secondary marriages with brothers' wives, and somewhat less so when sororal polygyny is prevalent and men cannot take brothers' wives.

Considering the extremes, then, asymmetry in cousin treatment should be concentrated in societies favoring or specifying sororal polygyny, and symmetry should be concentrated in societies practicing monogamy. This is precisely the case (figure 11). Thus, in Murdock's (1967) ethnographic sample of 565 societies, almost half (211) of the 423 societies for which relevant data are available treat parallel and cross-cousins symmetrically or do not distinguish them, and half (212) treat them asymmetrically or distinguish them. But 75 of 79 societies (95 percent) favoring or prescribing sororal polygyny treat parallel and cross-cousins asymmetrically, while only 35 of 101 monogamous societies (35 percent) do so (figure 11). The probability of this result occurring by chance is less than one in 10,000 ($p < 0.0001$). Alternatively, using Murdock's (1967) standard sample of 186 societies, substituting other appropriate societies

COUSIN TREATMENT	SOCIETIES		
	TOTAL	SORORALLY POLYGYNOUS	MONOGAMOUS
SYMMETRICAL	211	4	66
ASYMMETRICAL	212	75	35
TOTALS	**423**	**79**	**101**

Figure 11. The distribution of symmetrical and asymmetrical treatment of cousins in the two kinds of societies where they are expected to be most different if the asymmetry correlates with a probability that putative parallel cousins have a greater likelihood of being half siblings. The prediction is met.

where data are not available for a few societies indicated in the standard sample, one finds that only 5 of 15 monogamous societies (33 percent) treat cousins asymmetrically, while 7 of 8 sororally polygynous societies (87.5 percent) treat them asymmetrically (the probability of this occurring by chance alone is 1.77 percent using Fisher's exact probability test; Siegel, 1956). We do not know whether or not men are allowed to make secondary marriages with brother's wives in these cases because Murdock did not include this datum.

There are actually two kinds of symmetrical treatment of parallel and cross-cousins, those in which all are referred to as cousins and never classified as siblings (60 societies) and those in which all cousins and siblings are referred to by the same terms (132 societies, of which 23 are monogamous). In the latter case cousin marriages are obviously not allowed. Of 94 monogamous societies with the relevant data available, 73 (80 percent) disfavor or forbid all cousin marriages, so it is not surprising that 35 percent of the monogamous societies treating cousins symmetrically refer to them all as siblings (not as cousins) and forbid them to marry. The single exception to asymmetrical

treatment of cousins among the eight sororally polygynous societies in Murdock's standard sample, used above, is the Papago
Indians of Mexico (Underhill, 1939), who are described as dis-
favoring all cousin marriages and referring to both parallel and
cross-cousins as siblings (hence, this is not really an exception,
although I have so treated it in the test—see below).

So far I have mentioned two phenomena involving cousins in
which kinship terms referring to cousins seem not to reflect
genetic relationships: (1) parallel cousins are sometimes called
siblings, and (2) all cousins are sometimes called siblings. These
two departures from genetic kinship seem to reflect the two
general categories mentioned earlier, which together may explain a great deal about the background of human kinship
conventions. In the first case the apparent deviation is not a
deviation at all but a closer genetic fit than the observers had
thought. The second case appears at first to be an arbitrary
decision to serve some particular purpose. Such arbitrary decisions may be responsible for much of kinship nomenclature. In
this case, however, if the function of incest avoidance is served
by genetic outbreeding, then referring to all cousins as siblings
is not genetically arbitrary if it reflects a closer genetic relationship than in those cases in which cousins are not called siblings,
or if it leads to greater outbreeding in societies in which it is
advantageous to avoid first cousin marriages.

Other deviations of kinship terminology from genetic relationships are related to those discussed above. Thus, father's
brother and mother's sister are sometimes called father and
mother, respectively. As we have already seen, father's brother
may indeed be father, and mother's sister in sororal polygyny
will have offspring three-eighths like those of mother. Cross-
cousins are sometimes called by names applying to other generations such as aunt, uncle, niece, or nephew. This effectively
removes them from the marriageable category by referring to
them as relatives that are related by one-fourth and therefore
involved in incest taboos. Similarly, marriageable cross-cousins
may be called cousins, and nonmarriageable cross-cousins may
be called siblings. In each case we will wish to know eventually

if there are reasons for suspecting that nomenclatural practices deviating from those apparently genetically appropriate reflect historical variations in genetic differences not yet obvious to us.

Cousin Marriage Asymmetries and the Avunculate

Genetic asymmetries lead to conflicts of interest, in genetic terms, which we assume are expressed in phenotypic behavior insofar as it is possible for them to be identified and quantified as a consequence of historical consistencies in the pattern of social interactions among the individuals of the species involved. Within human families three main classes of potential conflict derive from genetic differences: parents versus offspring, husband versus wife, and sibling versus sibling. It is not necessary, in the arguments that follow, to quantify these different conflicts of interest. I assume that in the conflicts between parents and children parents are more likely to win, and in the conflict between spouses husbands are somewhat more likely to win. Both of these assumptions are based on the obvious power asymmetries.

More generally in society there are also conflicts between related and unrelated men, related and unrelated women. When the interests of men coincide and differ from those of women, we expect the men to cooperate to see their ends accomplished. Similarly, adult siblings of different sexes may cooperate against the interests of their respective spouses. Spouses may cooperate to see their interests realized against the interests of their offspring. Adults in different families may also cooperate to this end so long as the cooperation of adults from other families helps a parent to realize its own interests against those of its children and does not thwart the children's interests insofar as they conflict with the interests of members of other families or of society at large. I believe that variations in these kinds of power asymmetries, as a result of multiple environmental variables—taken together with the underlying genetic asymmetries within social groups—are the key to developing general theories about variations in cultural patterns.

Certain genetic asymmetries underlie the four kinds of cousin marriages (figure 10). For simplification I will make three assumptions, which can be varied as is useful in any further analysis. First, I will set confidence (probability) of maternity at 100 percent; hence, mother-offspring relatedness is always 0.5 in genes identical by immediate descent. Second, I will assume that initial spouses are unrelated, and, third, I will assume that the confidence (probability) of siblings that they share the same mother is 100 percent, and their confidence that they share the same father is the same as a man's confidence that his spouse's offspring are also his. These assumptions allow us to use a single figure for average relatedness of men to their spouse's offspring and siblings to one another. If we substitute X for this figure and Y for mother-offspring relatedness, the following formulas yield degrees of relatedness (in genes identical by immediate descent) between men and their grandchildren in the four kinds of cousin marriages:

1. Patrilateral parallel cousin marriages (a man marries his father's brother's daughter)

$$(X^2 + X^2Y + XY + X^3) / 2$$

2. Patrilateral cross-cousin marriages (a man marries his father's sister's daughter)

$$(X^2 + XY^2 + XY) / 2$$

3. Matrilateral cross-cousin marriages (a man marries his mother's brother's daughter)

$$(XY + X^2Y + X^2) / 2$$

4. Matrilateral parallel cousin marriages (a man marries his mother's sister's daughter)

$$(X^2 + XY) / 2$$

If confidence of paternity is set at 100 percent ($X=0.5$), 90 percent ($X=0.45$), and 80 percent ($X=0.40$) a man's relatedness to his grandchildren varies for the four different kinds of cousin marriages as in table 2.

TABLE 2
A Man's Relatedness to His Grandchildren in the
Four Kinds of Cousin Marriages

	Average Relatedness to Spouse's Offspring		
	0.500	0.450	0.400
Kind of Marriage	Average Relatedness to Grandchildren		
Patrilateral Parallel Cousins	0.375	0.310	0.252
Patrilateral Cross-Cousins	0.313	0.270	0.230
Matrilateral Cross-Cousins	0.313	0.264	0.220
Matrilateral Parallel Cousins	0.250	0.214	0.180

Note: Confidence of paternity varies from 100 percent to 80 percent; confidence of full sibship is regarded as equal to confidence of paternity; and confidence of maternity is regarded as 100 percent.

Marriages of cousins are most prevalent among the inter-mediate-sized societies of the world. Most of these societies are characterized by extended families or clans, and by a general lack of intense unity at the highest levels of association (Murdock, 1967; Hoebel, 1954). In other words, clans tend to be unified within themselves, and to be variously competitive and divisive with respect to one another even if their aggregate calls itself a nation and inspires some degree of national spirit, patriotism, or whole-group unity.

Many anthropologists have regarded the function of cousin marriages to be the establishment and maintenance of alliances, or subgroup unity (e.g., Levi-Strauss, 1969). Marriages lead to alliances—hence, whenever alliances are crucial, marriageable offspring are likely to be viewed as vehicles of alliance formation and maintenance, and those in power in any society may be expected to attempt to arrange marriages toward such ends and in their own interests.

Marriage patterns are also crucial because they determine the flow of inheritance (e.g., Goody, 1976). Of course, alliance and inheritance functions are affected by the nature of physiographic barriers between groups, kinds of heritable goods, and the genetic advantages and disadvantages of outbreeding, which are still largely unquantified.

Cousin marriages are arranged marriages. Their prevalence testifies to the general power asymmetry between parents and offspring, more specifically between coalitions of parents and their collective broods of offspring. That is, older people are in general more likely to control resources so that parents are in a position to engage in reciprocal altruism among themselves that enables each parent to realize his own interests against those of his own individual offspring when these differ.

If cousin marriages are to be the rule, then whenever it is profitable to have a few closely related descendants rather than many less closely related ones, a man may benefit if his children marry his own relatives rather than those of his spouse. This is so because he thereby increases his genetic relatedness to his grandchildren by 50 percent (in a first cousin marriage from one-fourth to as much as three-eights) without increasing the grandchildren's coefficients of inbreeding. This happens because he is related to grandchildren not only through his own offspring but also through his sibling's offspring. He accomplishes this increase in relatedness to his grandchildren at the expense of his spouse's ability to maximize her relatedness to her grandchildren. If one man realizes his genetic interest by causing his children to marry his sister's children, then obviously his sister's husband cannot enjoy this same advantage because his offspring will be married to his spouse's relatives. All men can realize their interests in this way only if all marriages are between patrilateral parallel cousins. Such marriages, then, might be expected to be most frequent in societies that are powerfully dominated by males, and in which heritable goods are relatively nonpartible, as with herds or land. Such marriages are the rule in Arab societies, in which male dominance is known to be extreme (Murdock, 1967), and in a few others.

Patrilateral parallel cousin marriages have been described as tending to keep herds and other goods intact and within the family, and this has in fact been given by participants in the system as the reason for their maintenance. On the other hand it has been argued by anthropologists that inheritance patterns are irrelevant, since (in some cases, at least) heritable goods are passed chiefly or solely to sons, so that they are kept together whether the son marries a relative or not. This view overlooks the crucial point that inheritance does not follow genes when children marry nonrelatives. Parallel cousin marriages cause inheritance and genes to stay together through both sexes of offspring even if only sons inherit directly, and patrilateral parallel cousin marriages benefit the collective of males more in this regard than any other kind of cousin marriage.

Similarly, there should be circumstances in which a woman will most benefit if her children marry offspring of her siblings, and the only way this can happen for all women is if all marriages are between matrilateral parallel cousins. Since these marriages would be most contrary to males' interests, and no societies are strongly dominated by women, we do not expect matrilateral parallel cousin marriages to be the rule in any society, and they are not. Even in matrilineal societies inheritance is actually passed from male to male *through* females (e.g., a man's heritable goods pass to his sister's son).

In most societies neither males nor females are completely dominant, and this may partly explain why parallel cousin marriages are so rarely the rule. One should, then, examine both the question of male dominance and that of confidence of paternity (see above) in attempting to assess the reasons for a particular attitude toward parallel cousin marriages, or frequency of them. One expects patrilateral parallel cousin marriages when male and female statuses are most different, and confidence of paternity is high, and cross-cousin marriages when confidence of paternity is low or when male and female statuses are most alike. One expects avoidance of siblings-in-law when parallel cousin marriages are the rule, and one does not expect parallel cousin marriages when philandering between siblings-in-law is permitted or prevalent.

In regard to asymmetries of relatedness to descendants, cross-cousin marriages introduce new complexities. If parallel cousin marriages cannot be arranged, or lead to deleterious inbreeding, then on the grounds given above, at least, a man is expected to prefer that his children marry his sister's children, and a woman that her children marry her brother's children. Each of these marriages will thwart not only the interests of the spouse, but those of the spouse's sibling and the sibling's spouse as well. If a man is able to arrange for his children to marry his sister's children, because the husband of his sister is automatically prevented from doing the same, alliances among the members of one sex against the members of the other sex are thwarted, but alliances between brothers and sisters against their spouses may be promoted.

Now we must ask how each parent should view the desirability of particular cross-cousin marriages of the two sexes of its offspring. If men successfully promote the marriages of their sons to sister's daughters, and/or women the marriage of their daughters to brother's son, the result is a patrilateral cross-cousin marriage system. If men successfully promote the marriages of their daughters to sister's sons, and/or women the marriages of their sons to brother's daughter, the result is a matrilateral cross-cousin marriage system.

Only if one sex of offspring marries the father's sibling's child and the other sex marries the mother's sibling's child can all of the members of either sex of parent be served approximately equally. When this happens, the system is either matrilateral or patrilateral with respect to cross-cousin marriages but not both. Thus, in a matrilateral system a man's sons marry his spouse's brother's daughters, but his daughters marry his sister's sons. This can be true for all men. In a patrilateral system a man's sons marry his sister's daughters, while his daughters marry his spouse's brother's offspring, and this can also be true for all men. Simultaneously, in a matrilateral cross-cousin marriage system a woman's son marries her brother's daughter while her daughter marries her spouse's sister's son. Again, both things can be true for all women. This symmetry may account in part for what Levi-Strauss saw as reciprocity and alliance maintenance between families, or exchange of women in connection with alliances between families. In part, it means that either matrilateral or patrilateral cousin marriage may be more likely than they would be otherwise in relation to systems in which both kinds are prevalent. Matrilateral cross-cousin marriage systems are much more frequent than patrilateral cross-cousin marriage systems, and we may now ask whether there is anything about the patterns of genetic relatedness that could contribute to this asymmetry.

Assuming men to be generally more powerful than women, if a man can only marry one sex of his offspring to his sister's children, because of resistance both from other men and from his wife, on genetic grounds which sex is he expected to choose? Or the question might be: from which sex of his offspring will he lose less by relinquishing their possibility of marrying his relatives rather than his spouse's relatives?

On average, a man will be more closely related to grandchildren through daughters because of the generally lower confidence of paternity as compared to maternity. This might mean that he would be more interested in investing in daughter's children. If males dominate, however, especially in a nonsororally polygynous society, investments intended for a daughter's children may not reach their goal but be diverted to the offspring of other wives unrelated to the investor. Even in monogamous or sororally polygnous societies in which males dominate, a man may gain considerably by controlling, guiding, or imparting status to his son-in-law. It would not be easy for a man to have similar positive effects through effort exerted on behalf of a daughter-in-law. Women usually have lower status than their husbands and are less appropriate objects of inheritance. Moreover, it would be difficult for a man to pay very much attention to a niece without arousing the suspicion of sexual interest. These last points were made by Homans and Schneider (1955) in arguing that marriage patterns depend upon who is "in authority" in each society or situation. As pointed out by Burling (1958) their explanation does not account for matrilateral cross-cousin marriages when the uncle of the groom is the "person of authority." I believe that the explanation I am giving here, which involves patterns of reproduction through the coincident flows of genes and inheritance, may account for this situation (see also Flinn, in press).

The above considerations suggest that men would be expected to prefer matrilateral cross-cousin marriages because they are slightly more closely related to daughter's children, because they are less able to influence the success of a son's children by helping a daughter-in-law, and if they are able to assist grandchildren through a daughter married to a nephew by assisting, guiding, and controlling the nephew (who may also be relatively closely related if confidence of paternity is low). In this system each young male will tend to be guided, controlled, and helped by both his father and his maternal uncle. Females will tend to be helped by their fathers and their husbands. In patrilateral cross-cousin marriages, on the other hand, males would tend to be helped by their fathers alone, females by their maternal uncles. Thus, it would appear that men have more balanced and extensive ways of assisting grandchildren under matrilateral as opposed to patrilateral cross-cousin marriages, and it would not matter whether the "person of authority" is a boy's father or his uncle. Flinn (in press) has shown that "patrilateral cross-cousin marriage, in almost all societies for which relevant data are available, is associated with high status father-son altruism in conflict with uterine kin altruism."

A final note about cousin marriages: among putative cousins unlikely to be half-siblings, patrilateral parallel cousins will tend to be most distantly related, because the parent of each cousin by which their relationship joins (their father-brothers) will each suffer some loss of confidence of parenthood. This means that on outbreeding grounds patrilateral parallel cousin marriages are the most likely of all cousin marriages; conversely, matrilateral parallel cousin marriages are least likely because offspring of full sisters are essentially certain to average one-eighth alike. In this regard cross-cousin marriages rank between the two parallel cousin marriages.

I believe that the appropriate analyses of genetic and power asymmetries, patterns of flow in genes and heritable goods, and sexual division of labor in societies with different economic bases will eventually lead to general explanations of the difficult problems of complex and diverse marriage patterns in human societies across the world. The principal ingredient here added to the anthropological discussions is that of genetic asymmetries. I make no pretense that the arguments given here explain the distribution and frequencies of cousin marriages and their entire relationship to the avunculate. I would only argue that genetic asymmetries like those I have just described should be a part of the understanding of anyone who undertakes analyses of these phenomena.

SOCIAL LEARNING AND CULTURAL PATTERNS

Matrilineal descent is normally linked with matrilocal residence, patrilineal with patrilocal.
—Linton, 1936, p. 169

> Both [neolocal and bilocal residence] are
> normally associated with bilateral de-
> scent, and. . . . The shift to neolocal
> residence results in the emergence of the
> isolated nuclear family; that to bilocal
> residence facilitates the development of bi-
> lateral kin groups and the bilocal ex-
> tended family.
> —Murdock, 1949, p. 208

I have argued that social learning may be the only general or widespread mechanism whereby individuals acquire the ability to behave in evolutionarily appropriate ways toward genetic relatives. I have shown that many patterns of culture accord with expectations from evolutionary theory. Now we can ask whether or not social learning situations exist in the appropriate asymmetries to yield the observed correlations between cultural patterns and evolutionary expectations. Many people will regard this question as crucial and in any case to answer it will complete the circle of connections between human behavior and evolution. For most situations the question is unexplored; here I will discuss it in only a preliminary way, in regard to a few simple cases.

The rewards and punishments for given actions are more obvious in some situations than in others. We can consider first some fairly simple and direct circumstances before taking up more difficult problems like the origin or maintenance of particular patterns of cousin treatment.

In every society in the world sexual relations between siblings or parents and offspring are forbidden, abhorred, and extremely rare (Murdock, 1949). There is evidence (Alexander and Noonan, ms.) that this response exists not only because parents and society in general forbid or disapprove of such behavior but also because the individuals that might be involved themselves avoid it. Few persons could describe the developmental or learning experiences that produce this avoidance in themselves but none can deny its universality throughout the world. The exceptions are always peculiar, such as brother-sister marriages allowed only in royal lines in ancient Egypt and Hawaii, or father-daughter sexual relations in certain

African tribes which are restricted to ceremonial situations associated with the father's departure on a hunt. As Murdock (1949) points out, these cases are so unusual as to support the proposition that incest avoidance is a universal tendency. We do not know all of the learning situations by which incest avoidance develops in the individual, nor do we understand precisely the functions of genetic outbreeding. We do know that among sexual organisms all but a few peculiar species consistently outbreed. In other words, we humans share the tendency to outbreed with nearly all sexual species. In this case, the pattern of our behavior is undeniable even though neither its function nor its ontogeny is known. Incest avoidance is a demonstration that outbreeding behavior exists in humans, and that it can have a cultural component even if we have no awareness from personal history of its basis and no real knowledge from either tradition or biology of its adaptive significance.

For other human behaviors all of these things do seem to be known. In a truck stop I once heard some drivers discussing an incident in which one of them had stopped to assist another driver. One driver remarked that after eating with another man and drinking coffee with him one could not pass by when he needed help. Another driver then said, "Well, you stop to help a man, he stops to help somebody else, and if you get in trouble somebody else stops to help you. And that's the whole ball of wax."

I have little doubt that truck drivers feel considerably more secure and less lonely when they know there is a strong tradition of mutual assistance on the highway. The good feeling that any of them gets upon stopping to help someone else is a reinforcement for behavior which they believe is increasing the likelihood of receiving help themselves when they need it. Much of the time their feelings in this regard are probably correct, especially when the assistance they give is relatively inexpensive and of great importance to the helped individual, and most especially when opportunities exist, like the one I observed, to perpetuate and enhance cooperativeness by discussing its virtues.

Here, then, is behavior in which the individual actors—or at least those like the man I overheard—for all practical purposes may understand rather well the reasons for their behavior. I believe that anyone can think of almost innumerable cases in which our behavior is guided in similar fashion—by what may be called "practical reason" of a very conscious sort. How, though, might the special treatment of sister's son in certain societies or the asymmetrical treatment of different kinds of cousins achieve and maintain a correspondence to adaptive predictions from biology? What might be the connections of the appearance and spread of such phenomena to variations in social learning situations?

In regard to treatment of parallel and cross-cousins and the outbreeding argument given earlier, intimate social interactions between an individual and spouse's same-sex sibling or same-sex sibling's spouse are more likely in polygynous than in monogamous societies. Within polygynous societies such relations are more likely in certain kinds of living arrangements than in others. Asymmetrical social interactions and greater likelihood of philandering or shifting of spouses between same-sex siblings are both more likely in the societies in which cousin treatment is most asymmetrical, and the social asymmetry is in the appropriate direction relative to the asymmetrical treatment of cousins. The philandering would lead to asymmetry in genetic relationships, and the social intimacy correlated with it would lead, through the kinds of learning differences postulated earlier, to a corresponding asymmetry in nepotistic favoritism and its well-established corollary in avoidance in marriage. The additional suggestion may be made that social relations appropriate to a given genetic relationship may be both promoted and readily accepted by those whom it serves.

Such circumstances follow almost too beautifully from the simple kinds of reinforcement or learning schedules that I have earlier postulated to lie behind the operation of inclusive-fitness-maximizing. I think we can be certain that the actual relationships are much more complicated. Nevertheless, this line of reasoning may open the way to understanding how cultural change can occur without genetic change, while nonetheless maintaining a structure consistent with the maximizing of inclusive fitness by individuals. The possibility of such social learning mechanisms weakens the contention that evolutionary approaches to human behavior are vulnerable because they cannot specify how reproductive maximization can be approached through cultural change. It removes any necessity of an intolerable determinism. It relieves the necessity of supposing that the evolution of culture and the evolution of genes must somehow be examined independently. The nature of the mechanism, moreover, blunts the criticism that the evolution of merely a capacity for culture makes the relationship between natural selection and the structure and variations of culture a triviality.

WHY IS INCEST SO ABHORRED?

In a 1978 televised interview with the anthropologist Marvin Harris, Dick Cavett remarked that no one had ever satisfactorily explained to him why humans should abhor incest so intensely. I would like to propose a partial answer to this question which bears on numerous topics discussed in this book, and also relates evolutionary biology to certain Freudian concepts (see also Alexander and Noonan, ms.).

Mating and offspring production with close relatives is very infrequent among organisms in general, whether animals or plants. In many cases there are obvious complex, evolved mechanisms preventing inbreeding. In others something as subtle as social avoidance or differential dispersal of the sexes eliminates or reduces inbreeding. Even though our understanding of the genetic consequences of inbreeding and outbreeding is still too incomplete to allow us to pinpoint with confidence the precise reproductive advantages of outbreeding, we can accept with confidence the notion that some degree of outbreeding is reproductively adaptive.

From this assumption we realize that incest will be disadvantageous not only to one's own reproduction but to the reproduction of all one's closest relatives as well, and both things are true whether the incest is committed by one's self or by one's close relatives. Incest is one of the few social acts for which this is so. It may be seen that this view of the results of incest, together with the near-universality and intensity of its abhorrence (in many societies it is punishable by death—Murdock, 1949), gives powerful support for the significance of the concept of inclusive fitness.

Now we must add a fact that makes incest paradoxical, as compared to other social acts. In general, multiplicity, intensity, and pleasurableness of social interactions with others of opposite sex represent the actual criteria upon which are based inclinations to proceed toward increasing intimacy, sexual behavior, and the production and shared tending of offspring. With regard to incest—and only incest—the precise opposite is true: those with whom we are most socially intimate and most pleasurably stimulated are the very individuals with whom sexual activities and marriage commitments would be most disadvantageous. It is thus not surprising that Freud and others should have regarded incest as "the core developmental problem" in human social behavior. It is not, however, that we have "powerful incestuous impulses" that must be overcome (e.g., Lindzey, 1967), but that we have powerful *reproductive* impulses, and that the very reinforcers by which we channel these reproductive impulses appropriately are necessarily pervasively involved in another set of interactions from which reproduction is maximized by an opposite outcome—by sexual avoidance rather than sexual intimacy.

In view of the great difficulty we have experienced in clarifying concepts of behavioral development, it is ironic that a set of learning experiences leading most of the time to highly reproductive behaviors should, in a single context, lead to unreproductive behaviors, and that the adjustment of human sociality around this conflict should yield a result that has to so many appeared as the one definite "instinct" or "inborn" human social tendency. Yet the startling fact that children reared together in the Israeli kibbutzim rarely marry (Spiro, 1958; Shepfer, 1978), and the evidence that the Taiwan practice of rearing bride and groom together as brother and sister leads to increased failures of marriage (Wolf, 1966, 1968)—despite contrary parental desires in both cases—seem to me to show beyond doubt that despite its universality incest avoidance is socially learned.

There is another irony: the more intensely an act is abhorred —presumably because of its deleterious effects on the inclusive fitness of the abhorring individual—the more likely it is that other individuals can manipulate and deceive around that act. The more intensely one wishes to avoid incest, the more possible it is that someone else can use this compulsion to his own ends, by striving to designate as incestuous those acts he does not wish the other (or his relatives) to commit. In other words we should expect that the rule-makers in any society will use the notion of incest avoidance to further the particular marriage arrangements that benefit them. In societies in which marriages are characteristically arranged by parents, it would not be surprising to find the label "incestuous" being attached to whatever marriages parents wish to discourage. The consequence would be that the cultural aspects or uses of the concept of incest (as opposed to the promotion of genetic outbreeding as such) will vary according to the history and the power structure of different societies. I believe that this source of confusion, around the central issues of incest and marriage, will turn out to be one of the most important apparent discrepancies between the biologists' and the social scientists' views of the patterns of culture.

RECIPROCITY AND NEPOTISM: SAHLINS' MODEL

The social anthropologist Marshall Sahlins provided in 1965 a review and analysis of the kinds of "reciprocity" occurring in "primitive" cultures. In 1975(a), I analyzed Sahlins' arguments in evolutionary terms, and in 1976(a), Sahlins, in a book critical of the evolutionary approach to human behavior, disagreed with my conclusions (see Alexander, 1977c, for a critical review of Sahlins' book). Because Sahlins is one of only a few vehement critics of the approach espoused here, and in his writings the most vehement of them all, I shall here recapitulate our exchange, following it with a critique of Sahlins' denial of my interpretation (see also Alexander, 1979c).

Sahlins (1965) described a "general model of the play of reciprocity" in primitive society by "superimposing the society's sectoral plan upon the reciprocity continuum" (figure 6). He divided reciprocity into three classes, which he terms generalized, balanced, and negative. " 'Generalized reciprocity' refers to transactions that are putatively altruistic. . . . The ideal type is Malinowski's 'pure gift' . . . 'sharing', 'hospitality', 'free gift', 'help', and 'generosity' . . . 'kinship dues', 'chiefly dues', and *noblesse oblige'.*" He mentions "the vagueness of the obligation to reciprocate" and uses as an example "for its logical value," the "suckling of children . . . the expectation of reciprocity is indefinite. . . . A good pragmatic indication of generalized reciprocity is a sustained one-way flow. Failure to reciprocate does not cause the giver of stuff to stop giving: the goods move one way, in favor of the have-not, for a very long period." (p. 147).

Sahlins' model concentrated generalized reciprocity in the household and implied its extension across the lineage sector of the village. He was in fact speaking largely of nepotism. Evolutionary biologists have come to expect that selection will mold organisms to assist their closer kin over their more distant kin, and kin over nonkin, even when reciprocity in kind is unlikely, and at least to concentrate such one-way beneficence on genetic relatives, perhaps dispensing it to no one else. Sahlins, although he would deny it vehemently, is actually telling the evolutionists that these expectations are fulfilled to an astonishing degree in primitive human societies in which kinship is "the organizing principle or idiom of most groups and social relations."

The major difference between Sahlins' generalized reciprocity and nepotism based on kin selection is that Sahlins has lumped sustained one-way flows of benefits dependent upon return in genetic currency (i.e., nepotism) together with other benefits which are really a form of balanced reciprocity. "Chiefly dues" and "noblesse oblige" are probably not best viewed as unreciprocated altruism, but rather as parts of exchange systems in which the returns are indirect or quite complex, or represent promises or obligations dependent upon future contingencies with certain probabilities of occurrence.

Balanced reciprocity, to Sahlins, refers to direct exchange.

> In precise balance, the reciprocation is the customary equivalent of the thing received and is without delay. . . . "Balanced reciprocity" may be more loosely applied to transactions which stipulate returns of commensurate worth or utility within a finite and narrow period. Much "gift-exchange", many "payments", much that goes under the ethnographic head of "trade" and plenty that is called "buying-selling" and involves "primitive money" belong in the genre of balanced reciprocity.
>
> Balanced reciprocity is less "personal" than generalized reciprocity . . . more "economic" . . . the pragmatic test of balanced reciprocity becomes an inability to tolerate one-way flows; the relations between people are disrupted by a failure to reciprocate within limited time and equivalence leeways. It is notable of the main run of generalized reciprocities that the material flow is sustained by prevailing social relations; whereas, for the main run of balanced exchange, social relations hinge on the material flow. [P. 148]

Sahlins' description tells us two important things about this class of reciprocal transactions. First, it is clear that returns to benefit-givers cannot be measured via gains in the phenotypes or reproduction of the recipients. If the system is to be maintained, reciprocation must accrue to the benefit-giver. One expects, although Sahlins does not mention the possibility, that reciprocation to one's close relatives might also satisfy such debts. In Sahlins' "balanced reciprocity" the actors are, in evolutionary terms, treating each other as if they were not genetic relatives. This suggests that the balance of genetic relatedness and reproductive competition among the participants is such that no gain is to be received by either person treating the other like a relative. Thus, reproductive competition between individuals generally drops with geographic distance, but so does genetic relatedness. Discussions of kin selection usually do not clearly specify this condition. Its results are that, as geographic distance from Ego increases, under certain conditions of gradual change in both relatedness and competitiveness, nepotism cannot evolve; and that, in any case, nepotism will cease to be favored at distances beyond which genetic relatedness continues to diminish (Alexander, 1974).

The second critical aspect of Sahlins' description is its emphasis upon the social relations of the participants. Repeatedly, he implies that participants in balanced reciprocity are associates and friends, i.e., individuals who expect to maintain good relations in the future and conduct their interactions accordingly. Cheating is minimal. It comes as no surprise that Sahlins sees balanced reciprocity as concentrated at the level of the tribal sector, and we may imply that it extends inwardly into the village sector, and probably into the lineage sector, but not, although Sahlins does not specify this point, beyond tribal limits, except when coalitions, temporary or otherwise, cause tribal limits to be rather indefinite.

Again, Sahlins' model is an uncanny match for that of the evolutionary biologist. Indeed, these first two kinds of trading or beneficence are roughly what the biologists have called kin selection (Hamilton, 1963, 1964, 1972; Maynard-Smith, 1964) and reciprocal altruism (Trivers, 1971), respectively. Evolutionary theory not only recognizes the two categories almost precisely as Sahlins has distinguished them, with the exceptions noted, but it specifies the social levels at which generalized reciprocity gives way to balanced reciprocity (figure 6).

Sahlins' emphasis on the continuation of social relations among participants in balanced reciprocity raises the question of why this should be so. Why should bartering individuals maintain a balance when they have opportunities to cheat, seemingly to their own advantage? But let us delay this question until we have considered Sahlins' third category, which he described as follows:

"Negative reciprocity" is the attempt to get something for nothing with impunity . . . "haggling", "barter", "gambling", chicanery", "theft", and other varieties of seizure . . . the aim . . . is the unearned increment. One of the most sociable forms, leaning toward balance, is haggling conducted in the spirit of "what the traffic will bear." From this, negative reciprocity ranges through various degrees of cunning, guile, stealth, and violence to the finesse of a well-conducted horse-raid. The "reciprocity" is, of course, conditional again, a matter of defense of self-interest. So, the flow may be one-way once more, reciprocation contingent upon mustering countervailing pressure or guile. [Pp. 148–49]

Negative reciprocity, as one may predict, centers outside tribal and national boundaries. The more overt or blatant the cheating, the less likely one is to conduct it among relatives or groups of friendly persons, and the more likely he is to receive admiration or appreciation for success involving strangers or, better yet, mutual enemies.

Sahlins (1976a), in a book denouncing the use of evolutionary theory in the analysis of human sociality, has denied the correspondence I suggested between his model and predictions from evolutionary theory, in the following words:

. . . sociobiologists, notably Alexander (1975), have taken the equally well-known tendency of economic reciprocity to vary in sociability with "kinship distance" as cultural evidence of biological "nepotism," hence as a proof of kin selection. . . . this conclusion is based on an elementary misunderstanding of the ethnography. The kinship sectors of "near" and "distant," such as "own lineage" vs. "other lineage," upon which reciprocity is predicated, do not correspond to coefficient of relationship, so the evidence cited in support of kin selection (e.g., Sahlins, 1965) in fact contradicts it. . . . [P. 112]

Sahlins' denial seems to me preposterously careless and incomplete. He does not explain why his detailed descriptions of the three kinds of reciprocity are almost perfect reflections of evolutionary analyses of nepotism and reciprocity (Trivers,

1971; Alexander, 1974, 1975a; others); or why, within the social system, they should center so precisely where evolutionary theory predicts they will. His denial is focused on the nepotistic aspects of my analysis. But the example of "suckling children" is his own. Is he denying that people are altruistic toward their offspring, or that one-way flows of benefits typify such altruism? The word "lineage" is his own. Does he deny that persons in the same "lineage"—explicitly as anthropologists use the term—are more often genetically related, or more closely related, than persons in different lineages? Murdock (1949) could scarcely have been more explicit on this point. After noting that "descent refers only to social allocation and has fundamentally nothing to do with genealogical relationships or the recognition thereof" (p. 15) he states that "a consanguineal kin group produced by either rule of unilinear descent is technically known as a *lineage* when it includes only persons who can actually trace their common relationship through a specific series of remembered genealogical links in the prevailing line of descent" (p. 46). At least in part Sahlins evidently refers to what anthropologists term the "classificatory" aspects of kinship systems. I have discussed these earlier, in biological terms, and they appear in no way to negate the analysis given here.

APPARENTLY NONREPRODUCTIVE BEHAVIORS

It is paradoxical that when evolutionary analyses of human behavior are discussed their proponents are likely to encounter two quite different responses: first, that such analyses cannot explain this or that—adoptions, suicide, homosexuality, asceticism, or others—and, second, an exasperation that evolutionists seem to think they can explain everything.

Steven Gould (1978), in a critical review of Wilson's *On Human Nature* (1978), claims that we cannot test evolutionary hypotheses—which he says all too often take the form of "just-so stories"—because a "genetic story" of some sort can be made up as a plausible explanation of almost anything.

But evolutionary hypotheses can be tested, and, of course,

they only become useful when they become testable. Means of falsification exist in many cases, and in others deviations from expectations on the basis of chance can be shown to be highly significant. Both points are demonstrated in this book.

But what about topics like adoption, homosexuality, asceticism, and suicide? First, we should distinguish two possible situations. One is that no plausible explanation or hypothesis, consistent with biology, exists for any of them. The alternative is that several or many possible explanations exist, and we have not yet been able to choose among these various alternatives—or, on the other hand, to eliminate them all. If the first situation were true for any behavior a definite threat would be posed for evolutionary explanations. The second situation, however, poses no such threat, even if the possible explanations are in fact still in the form of "just-so" stories. What we must do is change just-so stories into scientific hypotheses and then test them.

I shall not attempt here to analyze in depth any of these seemingly paradoxical behaviors. But I will show that plausible explanations exist that are consistent with evolutionary theory, that definite reasons exist for regarding them as plausible, and that testable hypotheses can be generated around them.

First, I do not necessarily suppose that adoption, homosexuality, asceticism, and suicide are singular topics with single explanations. Rather I would suggest that in each case a reasonable approach is first to construct a list of all of the apparently different "kinds" of adoption, homosexuality, asceticism, and suicide. If, in each case, all of the different kinds turn out to have common elements—or even a common explanation—then so much the better; but this need not be the case.

Second, I would attempt to describe the specific circumstances under which each of these behaviors occurs; none of them is characteristic of a majority of any population, and each appears connectable to special situations.

Third, I would ask if the situations in which these behaviors tend to occur most often are such that the behavior is unusually likely to lead to reproduction via an unusual route appropriate to that situation. For example, it would surely occur to me that life insurance companies commonly include in their contracts a refusal to pay one's beneficiaries (usually a spouse or relative) if one commits suicide; or, if such a clause is not included, the premium is likely to be much higher. I might ask if the phenotypes of some homosexual individuals, or their developmental experiences, are such as to reduce their likelihood of success in ordinary sexual competition or increase their likelihood of success in some alternative kind of behavior. I would ask in what kinds of societies asceticism has been most prominent, which individuals are most likely to adopt such behavior, and what are the effects of their behavior upon close relatives.

In the case of adoptions I would surely look directly to those cultures—for example, in Oceania (Carroll, 1970) and the Arctic (Burch, 1975), where adoptions are uniquely prevalent, and search for correlates both common to all such societies and unique among them. I would notice immediately that these are societies in which extraordinary importance is attached to reciprocity—to commitment to the assistance of one's neighbors and associates. I would be aware that another kind of extreme behavior of a parallel sort—allowing another man sexual access to one's wife—also has been widely reported in one of these kinds of societies (certain Eskimos—Burch, 1975).

All of these are mere introductory suggestions. But they are the kind that one must continue to pursue if he is to expect to test the questions raised by *apparently* nonreproductive behavior of all sorts in human societies. Obviously, it is not useful to claim that explanations consistent with evolution are already known for every such case, when these "explanations" are in fact only tentative hypotheses, in some cases with many possible alternatives, and sometimes still in untestable forms. On the other hand, considering extreme topics of this sort in even a cursory fashion emphasizes that no behaviors are yet known that represent significant threats to the general importance of evolution as a force underlying all of human behavior.

To conclude these remarks I will cite several passages by Burch (1975) and Carroll (1970). Neither author wrote with an evolutionary hypothesis in mind, and they were writing about behaviors that seem biologically paradoxical; yet I contend that one could scarcely hope for accounts more supportive of the approach I am suggesting here. First, Burch:

> The most fundamental consideration in traditional Northwest Alaskan Eskimo strategies of affiliation was that not a single goal in life, including the basic one of sheer survival, could be achieved without the help of kinsmen. An individual could relate peacefully to non-kin, for example, in terms of a partnership . . . but one could never depend on a partner who did not have the support of several relatives himself. Since this was a reciprocal consideration, it follows that people had to be actively affiliated with kin in order to participate successfully even in this non-kin relationship. But even the strongest non-kin tie was considered weaker than the weakest kin relationship. In times of crisis, such as famine or war, one always had to opt for a kinsman in favor of a partner, and one knew that one's partner would have to do likewise. There was practical as well as moral compulsion in this, because, *if* an individual *failed* to support a relative over someone else, he would be ostracized summarily by all the kin who knew about his behavior [p. 198].

> The primary form of non-residential marriage among the traditional Northwest Alaskan Eskimos was what I refer to as "co-marriage," but which is known elsewhere as "wife-exchange," "wife-trading," and "exchange-marriage," to name only the most common labels. It consisted of the union of two conjugal (i.e., resident husband-wife) pairs into a larger marital unit via the mechanism of sexual intercourse between each man and the other's wife. . . . There is reason to believe that the basic emphasis then may have been on co-marriages between, rather than within societies, this institution being one of the most effective alliance mechanisms the Eskimos had . . . once the union had been established, it continued for the lifetime of the members, regardless of whether or not sexual relations that established it were ever repeated. . . . my informants . . . pointed out that in the old days, a person without kinsmen was a person whose days were numbered; the people in the traditional period were acutely aware of

this fact. Co-marriage was just about the best way in which people could increase the number of their relatives in a short period of time [pp. 106, 108–9].

Adoption was widespread among the traditional Northwest Alaskan Eskimos, and it was still relatively common during the recent period. . . . In traditional times particularly, the practice functioned to distribute children in such a way that the mean, the median, and the mode figures for the number of children living with each conjugal pair were probably virtually identical, namely two.

When an Eskimo child was adopted, it became a full-fledged member of the adopting family. At the same time it continued to retain *all* its connections with the family into which it was born. Incest taboos applied to both sets of kin. For these reasons, a child adopted out became an "other-family" offspring of the donor family. The child would have two sets of parents, a real set of "other-family" parents and an adopting set of "same-family" parents.

The Eskimos commonly adopted children already related to them. Until well into the recent period, they *always* adopted the offspring of people *known* to them. Most frequent were adoptions by grandparents in the Eskimo sense of that term, i.e., by consanguineal relatives of the second ascending generation (regardless of whether they were lineal kin or not). The most common pattern seems to have been for relatively young grandparents whose own children were at or approaching maturity to adopt their first grandchild, the child simply staying with them when its parents decided to move out of their household. Adoption by relatives other than grandparents was also common, particularly by same-generation consanguines who apparently were not going to have any living children of their own.

A child who was adopted by people who were already related to him became connected to them in terms of two (or more) relationships simultaneously. As far as the principles were concerned, the initial connection was always replaced by the appropriate parent-child relationships, thus strengthening the tie between the individuals concerned . . . stronger relationships were substituted for weaker ones; that is, people changed from cousins to siblings, and from uncles to fathers, rather than *vice versa* [pp. 129–30].

Paralleling Burch's writings are the statements of the cultural anthropologist Vern Carroll, in the remarkable volume, *Adoption in Eastern Oceania* (1970), which he edited. After noting that even in the United States about half of the adoptions are by relatives, he continues:

> Whereas American adoption is often a transaction involving total strangers, adoption in Oceania is generally a transaction between close relatives. In the typical case, the adopter is related to one of the natural parents of his or her adopted child as a full or classificatory "sibling" or "parent" [p. 5]. . . . Child-adoption appears to be only one of many possible ways of establishing *as if* kin relationships. . . . Adoption, of whatever sort and however defined, establishes long-term, solidary relationships of the kind which are usually ascribed culturally to close biological kin [p. 10]. . . . It is difficult to refuse a relative's request for a child— especially if the request is made before the child's birth—since to do so is to deny the existence of a bond of kinship with the person who is making the request. Relatives should be willing to share all of their resources, including children, especially when one of the relatives is in need. . . . The closest analogue in our society to a Nukuoro adoption request would be an invitation to a dinner party: to refuse for no reason is to risk rupturing whatever relationship has been established [p. 125]. . . . it is clear that the fundamental relationship in adoption is the close sanguineal relationship between *one* of the child's natural parents and *one* of the adopters [p. 128]. . . . Nukuoro children move back and forth freely from the household of one relative to that of another; even so they are rarely more than a few hundred yards from their natal home. . . . To a certain extent *all* of a child's senior relatives share responsibility for his care. . . . Under no circumstances will "strangers" be permitted to adopt a child [p. 130]. . . . Not giving up children to "strangers" is simply a refusal to accept these strangers as kin. . . . Asking for a child and agreeing to an adoption are expressions of appropriate kin sentiment which give momentary pleasure to both parties. . . . Many adoption requests are made impulsively, often when it is first noted that a female relative (or a male relative's wife) is pregnant. Such requests appear to be motivated, at least in part, by a desire to "test" the relationship —to see if the individual who is asked is really a "relative" and

will act as a relative should by agreeing to the adoption [p. 131].
. . . Kinship on Nukuoro implies a host of vaguely defined reciprocal obligations, the most important of which perhaps is the obligation to share. One should always be willing to assist a kinsman.
. . . The ethics of sharing is continually reinforced by offers to
share, even when there is no apparent need [p. 146]. . . . the
repeated experience of adoption must communicate some sort of
information to the individual about the premises of his culture.
My view is that the most important information which is encoded
in this sort of life experience is that relatives should share with
each other. Any specific act of adoption encodes a number—
probably a very large number—of messages about the relationship of the principals, but it is here maintained that the cultural
message which all adoption acts communicate to all of the parties
concerned is that relatives are interdependent and that the maintenance of this network of interdependency must take priority
over the wishes of individuals, even such strong wishes as attach
to one's natural children [p. 152].

It seems reasonable to hypothesize that in the Eskimo and
Oceanic societies studied by Burch and Carroll, group relationships are so important to survival and reproductive success that
unusual systems of alliance have generated. These systems have
evidently developed out of previously existing patterns of nepotism and marriage, and they have come to involve complex but
definite interactions of nepotism and reciprocity, including
adoptions and co-marriages.

THE BIOLOGICAL DISTINCTIVENESS OF THE HUMAN SPECIES

Anthropologists have long been interested in identifying
those traits that are peculiarly human, such as culture, consciousness, tool use, language, and what Leslie White called
"symbolling." Similarly, anthropologists have not infrequently
attempted some kind of general comparisons of patterns in
human societies, along the lines of that by Flannery (1972)
illustrated in figure 12. Here I would like to consider both of
these topics briefly from the viewpoint of modern evolutionary
biology.

Why Are Humans Different?

Alexander et al. (1979) described reasons for assuming, from its current attributes, that the human species has been polygynous during much of its recent evolutionary history—that is, that, generally speaking, fewer males than females have contributed genetically to each generation, although harems have not necessarily been involved. Elsewhere in Chagnon and Irons (1979) many of the consequences of this conclusion were described, and indeed confirming findings. Considerable evidence also indicates that humans have essentially always lived in bands of close kin, probably containing more than a single adult male (e.g., Lee and DeVore, 1968). These two characteristics, however, fit a large number of nonhuman primate species. Alone they tell us nothing about how the human species came to possess its numerous distinctive attributes.

Alexander and Noonan (1979), in an effort to analyze the background of the concealment of ovulation in the human female, undertook to list the distinctive features of humans. They began with the usual list, including the above items:

1. Consciousness (self-awareness)
2. Foresight (deliberate planning, hope, purpose, death-awareness)
3. Facility in the development and use of tools (implying consciousness and foresight)
4. Facility in the use of language and symbols in communication (implying consciousness and foresight)
5. Culture (a cumulative body of traditionally transmitted learning—including language and tools, and involving the use of consciousness and foresight)

As suggested by the parenthetical comments, these five attributes are closely related to one another, perhaps inseparable. They are also not strictly comparable to one another: thus, consciousness and foresight are aspects of the human *capacity* for culture, while language and tools are *vehicles* and *aspects* of culture. Culture, in turn, by its existence and nature, and through its changes, becomes a central aspect of the environment in which the capacities of individuals to acquire and use consciousness, foresight, and facility in the development and use of language and tools have been selected.

Although these five attributes may once have been regarded as uniquely human, it now seems likely that all occur in other primate species, and chimpanzees alone may possess all five, though not in the form or to the degree that they are expressed in humans (Lawick-Goodall, 1967; Gallup, 1970; Premack, 1971; Fouts, 1973; Rumbaugh et al., 1973; Gardner and Gardner, 1969, 1971; Mason, 1976). To understand this set of related attributes, and why humans possess them, we must determine their relationship to the reproductive success of individuals during human history, when the capacities and tendencies to express them were originating and being elaborated. Despite the attention paid to them, these attributes have not been extensively analyzed as contributors to reproductive success.

Numerous other traits are also distinctive to humans (d, in the list below) or distinctively expressed in humans as compared to their primate relatives (de). They may be either cultural (c) or noncultural (nc) in origin, and universal (u) or not universal (nu) among humans. Surprisingly, only the first of these attributes (6) is evidently sexually symmetrical (s) in its expression. The rest are sexually asymmetrical (as) or involve (1) interactions of the sexes in connection with parental care (especially 9–16) and (2) group-living (especially 17–30).

6. Upright locomotion usual (de, nc, u, s)

7. Frontal copulation usual (de, c & nc, u, as)

8. Relative hairlessness (de, nc, u, as)

9. Longer juvenile life (d, nc, u, as)

10. Greater infantile helplessness (de, nc, us, as)

11. Parental care frequently extending into and even across the offspring's adult life (d, c, u?, as)

12. Unusually extensive parental care for a group-living primate (de, c?, nu?, as)

13. Concealed ovulation in females (sometimes described as continuous sexual receptivity, continuous oestrus, "sham" oestrus, or lack of oestrus (d, nc, u, as)

14. Greater prominence of female orgasm (perhaps—but see Lancaster, 1979) (d, nc?, u, as)

15. Unusually copious menstrual discharge (d, nc, u, as)

16. Menopause (d, nc, u, as)

17. Close association of close kin of both sexes, sometimes throughout adulthood (de, c, u?, as)

18. Extensive extrafamilial nepotism (de, c, u?, as)

19. Extensive extrafamilial mating restrictions (d, c, u, as)

20. Socially imposed monogamy (d, c, nu, as)

21. Extreme flexibility in rates of forming and dissolving coalitions (d, c, u, as)

22. Systems of laws imposed by the many (or powerful) against the few (or weak)(d, c, nu, as)

23. Extensive, organized, intergroup aggression; war (d, c, u?, as)

24. Group-against-group competition in play (d, c, u?, as)

25. Ancestor worship (d, c, nu, as)

26. Political and other kinds of appointed, elected, or hereditarily succeeding leaders (d, c, nu, as)

27. The concepts of gods and life-after-death (d, c, nu, as)

28. Organized religion (d, c, nu, as)

29. Nationalism; patriotism (d, c, nu, as)

30. Nations of thousands or millions of nuclear families (d, c, nu, s)

Concealment of Ovulation and Parental Care

Noonan and I reasoned that if we could explain certain of the more distinctive of these attributes, the backgrounds of some of the others might also fall into place. We were especially interested in menopause and the concealment of ovulation because of their uniqueness to human females, because they are physiological as well as behavioral traits, and because of their possible relationship to some of the other attributes. Menopause had already been seen as an aspect of parental care, diverting the human female from continued production of offspring—in humans requiring parental care for as long as fifteen or twenty years—to a tending of her existing offspring (Williams, 1957; Alexander, 1974; Dawkins, 1976). We reasoned that concealment of ovulation might also have to do with the unique extensiveness of human parental care, and therefore the unique helplessness of the human juvenile and length of the juvenile period.

Ultimately, Noonan and I concluded that concealment of ovulation evolved in humans because it enabled females to force desirable males into consort relationships long enough to reduce their likelihood of success in seeking other matings, and simultaneously raised the male's confidence of paternity by failing to inform other, potentially competing males of the timing of ovulation. In effect, concealment of ovulation disenfranchised those males who devoted themselves to monopolizing females during ovulation (and helping or protecting them only then) and increased the reproductive possibilities of males willing to remain with a female, protecting and helping her for a longer period and (eventually) capable of assuring her of assistance in the rearing of offspring if she in turn provided confidence of paternity. We argued that this could happen only in a group-living situation in which alternative mates were readily available, and in which the importance of parental care to the reproductive success of offspring was increasing, and that these two circumstances together describe a significant part of the uniqueness of the social environment of humans during their divergence from other primates.

Next we asked what challenges, in the form of differential reproduction in a group-living species, might have caused the emphasis on (1) consciousness and foresight, (2) social living, and (3) parental activities, revealed in the above list of human attributes. If we assumed that human attributes arose as a result of changes in physical or nonhuman biotic selective forces, such as climate, predators, or food shortages, then an evolutionary sequence diverging humans so far from other species in the particular directions they have taken, with all the intermediate stages becoming extinct, seemed to us difficult to reconstruct (see also Alexander, 1971). In the absence of any clear evidence of a massively unique selective environment in these respects for humans, it would be necessary to postulate that our uniqueness as a prehuman primate preadapted us to respond uniquely to some not-so-unique concatenation of environmental conditions, thereby evolving humanness. We regarded this hypothesis as unlikely.

The alternative, we reasoned, is that something about the evolving human species itself explains the differential reproduction that led to the divergence of the human line, and the extinction of close relatives along the way. The attribute that could explain human uniqueness, we argued, was an increasing prominence of direct intergroup competition, leading to an overriding significance in balances of power among competing social groups, in which social cooperativeness and eventually culture became the chief vehicle of competition. The probable relevance of complex social competition to human intelligence, consciousness, and foresight—and elaborate social tendencies —had been emphasized before (Darwin, 1871; Keith, 1949; Fisher, 1930; Alexander 1971–77; Bigelow, 1969; Alexander and Tinkle, 1968; Carneiro, 1970; Flannery, 1972; Wilson, 1973a). Intergroup competition had not, however, been linked to the human emphasis on parental care or the unusual sexual attributes of humans.

There is evidence of widespread dispersal and interaction of human groups distinct enough to be separated through fossilized fragments, during the long period of human genetic evolution (Coon, 1963). This evidence implies, at least, that countless interactions must have occurred between groups that were less distinct; recent human history supports this view and suggests that such interactions would often have been aggressive or competitive and that their outcomes were crucial in determining the later distribution and nature of the human species.

If intergroup competition was a principal guiding force in human evolution and if groups as a result grew in size and in social unity among the individuals comprising them, then parental care could have increased in value in two general ways. First, larger group sizes inevitably meant intensified competition for resources within groups. Juveniles, lacking the strength and sophistication to compete successfully for themselves, would have benefited increasingly from parental protection and assistance in securing resources for growth and future reproduction. Second, the intensification of competition both within and between groups would imply a growing conflict in individual reproductive strategies by pitting the value of direct competition with other group members against potential gain from their cooperation in intergroup competition (or within-group alliances). Individual reproductive success would depend increasingly on making the right decisions in complex social situations involving self, relatives, friends, and enemies. Critical choices would be aided by experience, and an intimate knowledge of the particular social environment. Parents who could impart this information to their offspring and the social skills for using it and expanding it, while providing guidance during the vulnerable years of learning, would have realized increasing reproductive advantages over parents who failed to so equip their offspring. Buffered against physical and social disaster by parental protection, the human juvenile evolved to abandon efforts at serious direct competition, becoming increasingly helpless over longer periods, while evolving extraordinary abilities to absorb and retain information and develop skills through attachment, identification, imitation, and more formal learning in early years. In other words, because of the existence of groups intensely competitive against one another, and because of the complexity of social competition within groups evolving to be effective in intergroup competition, the human species in some sense became its own most important selective environment, and the pressure of evolutionary change focused increasingly on parental care. The neotenous trend thus set into motion long ago seems to be paralleled in modern Western societies during difficult times by the extended preparations for professional life by children of parents who are sufficiently affluent to assure their offspring of a college education, by the added importance of advanced degrees when jobs are scarce, and in the extreme by additional training, as through postdoctoral training in the academic world. Currently, in the United

States, many so-called academic and professional people do not actually compete in their job markets until the age of thirty or beyond. This extraordinary trend is a consequence of changes in likelihood of success resulting from increases in social competition within human society, and it is evidence that under intense social competition within human groups, extensions of parental care are regarded as useful in improving one's chances of success. Many students secure most or all of their sustenance not from parents but through various surrogates of parental activity provided by the society at large (for example, food stamps, federally funded work-study programs, and governmental fellowships). Nevertheless, the principle seems clear.

Among early humans, tendencies to commit infanticide on the offspring of other males, or force the desertion of such infants, would have benefited males whenever such practices hastened ovulation in females and preserved female reproductive effort for the male's own offspring. The increasingly intense parental care of human mothers, its greater duration, and the wider spacing of babies associated with more intensive early parental care and infant helplessness would have enhanced the benefits of infanticide and enforced desertion of children to males acquiring females from other males. Thus, an important early aspect of male parental care may have been protection of the child against other group males competing for the female as a reproductive resource, and a crucial aspect of female parental care would have involved securing the protection of offspring by a capable male. Observations by Bygott (1972) suggest that strange chimpanzee infants (fatherless in their adopted groups) are vulnerable to infanticide by males within the group. Accounts of men (or women) killing children made fatherless by intercommunity warfare and exchange of women, such as among the Yanomamö Indians of South America (Biocca, 1971), imply that this may have been an important selective context for parental care in human history as well, and another indirect consequence of the extension of juvenile dependence.

In ancestral humans, then, an orphan, even at an advanced juvenile stage, was probably doomed to social impotence and

reproductive failure, if not prereproductive death, unless it was
a female old enough to interest a mature male; the extremely
derogatory connotation of words for "fatherless" juveniles in
nontechnological societies supports this inference. On the other
hand, juveniles with powerful parents and other relatives must
have been essentially certain of high success. Indeed, the un-
stratified or egalitarian bands presumed to represent the ancestral kind of human sociality can almost be defined by saying
that in them the major resource by which reproductive competition could be maximized is kinsmen.

Thus it appears that many of the unique attributes of the
human species can be understood by linking the extensive parental care of humans to the reasons for their ever-increasing
tendencies, throughout history, to live in larger, more complex
social groups, discussed in detail in the next chapter in connection with the rise of law and changes in the concepts of justice
and morality.

BIOLOGY AND THE LONG-TERM PATTERNING OF HUMAN HISTORY

Although it may be possible to develop a list of distinctive
attributes of humans, compared to nonhuman primates, within
recorded history human culture has been represented by an
almost bewildering diversity of patterns and practices. The real
question is whether or not arguments like those I have been
making here, from the principles of organic evolution, can help
point the way to the generalizations that have so far largely
eluded students of human culture and its history. I believe that
they can, and that we are on the verge of reconstructions and
explanations of the long-term natural history of humans that
will make sense, not only to both social and biological scientists, but as well to thoughtful people in all walks of life. I think
we are ready to move from the piecemeal, item-by-item kind
of analysis of culture that I have had to employ in this chapter
to broad-scale and comprehensive interpretations that will provide insight into countless features of human existence.

Here I will make only two comments on this general problem.

First, it is common to hear the remark, among social scientists, that evolutionary biologists who wish to analyze culture are trying to explain a variable with a constant. The variable is culture, and the "constant" is the principle of evolution by natural selection. People who argue this way must suppose either that there is no biological basis for variability in behavior or that when behavioral variations are heritable through learning they are unrelated to genetic adaptation. I believe that both arguments have been disposed of earlier in this book. I think we have already demonstrated beyond doubt that it is reasonable to expect that much of the variation in human behavior and culture reflects our inevitable background of differential genetic reproduction, leading to the elaboration by natural selection of the capability of responding in reproductively appropriate fashions to an extraordinarily wide array of social and physical environments. Even if we are yet unable to specify the proximate mechanisms whereby cultural variations can reflect a history of differential genetic reproduction, the best hypothesis at the moment is that such mechanisms, connecting culture and genes, do nevertheless exist.

The second comment involves the reconstruction by Flannery (1972) of "the evolution of civilization," or the rise of nation-states (figure 12). This diagram is particularly fascinating to biologists because it hints at just the kind of overview of cultural history from which general theories or models of our long-term history might be developed. In the next chapter I discuss Flannery's reconstruction in some detail because of its relevance to the rise of law. Here I will only note that the cultural institutions it compares differ from those that might enter the mind of a biologist making a similar effort. In figure 13 I have reconstructed Flannery's diagram so as to include a different set of cultural traits, some of which are more closely allied to topics a biologist might initially consider. I found it difficult to remain general, yet produce a single pattern varying coherently with the classes of societies used by Flannery. Perhaps this means that different diagrams will have to be constructed for different parts of the world, where cultural changes

—particularly among so-called tribes and chiefdoms—have proceeded independently under different ecological circumstances and from somewhat different earlier stages.

For example, arranged marriages between cousins are not typical of all societies falling into the categories of tribes and chiefdoms, although they are concentrated there; and they appear to be more typical of African societies than, say, those of the New World.

Perhaps the difficulty of remaining general while comparing biological features of societies is a commentary on the argument that reconstructions of this sort are too "unilineal" in their representation of cultural history. Perhaps the topics of interest to biologists are somewhat less general than those to which anthropologists look first. Perhaps a closer look at the original data, or the societies themselves, will reveal errors and misinterpretations that clarify the difficulties. Perhaps the biological-evolutionary view of cultural history will in the end turn out not to be as useful as some of us expect.

Whatever the case, the kind of reconstructions attempted in figures 12 and 13 seem to me to be what lies ahead along the route I have attempted to describe and justify here, in the further analysis of culture. The true excitement and magnificence of this aspect of human natural history will come from the eventual melding of what is known of the long-term geographic and temporal history of our various physically different ancestors, and relatives, with the more recent history of the distribution and nature of our cultural patterns across the face of the planet.

4

Evolution, Law, and Justice

===========

INTRODUCTION

The only general conclusion to be drawn is that, in any society that preserves a modicum of individual responsibility, there is a tension between individual ethics and social morality on the one part, and social morality and the legal order on the other part. How much these three spheres of normative order influence and modify each other is a question that cannot be answered in absolute terms.
                    —*Friedmann, 1967, p. 47*

So far I have talked about evolutionary analyses of human behavior strictly in historical terms: How have people unaware of the details of the process of natural selection behaved in relation to their relatives and nonrelatives? Do the data amassed by social scientists about human social and cultural patterns meet the predictions of Darwinian selection? I think that we have seen enough positive results to merit asking the next set of questions, which bear upon whether or not a Darwinian view of human behavior can help us interpret systems of laws, concepts of justice, and ethical questions. Will this approach im-

prove our understanding of existing systems of laws and ethics
—even the activity called "science," supposedly involved in the
development of the ideas and arguments in this book? And,
finally, what role might an evolutionary understanding of
human behavior play in the establishment, evaluation, and
maintenance of the rules of the future—the norms and ethical
decisions of people who understand their backgrounds more
thoroughly than anyone ever has before?

I wish to discuss separately the two issues of (1) the biologi-
cal-historical background of ethics and morality (the "is" and
"was"), and (2) the use of such knowledge in designing society
or controlling behavior (the "will be" and "ought to be"). My
conclusions on how biology affects these two issues will differ:
I will argue, and I hope show, that evolutionary analysis can tell
us much about our history and the existing systems of laws and
norms, and also about how to achieve any goals deemed desirable; but that it has essentially nothing to say about what goals
are desirable, or the directions in which laws and norms should
be modified in the future.

Causes of Human Groupings

Intergroup antagonism is thus the inevi-
table concomitant and counterpart of in-
group solidarity.
—*George Peter Murdock, 1949, p. 83*

Laws are only made by humans living in the kinds of social
groups we call societies. As a result understanding the particular
reasons for humans' having evidently always lived in social
groups composed of several or many families is evidently
closely related to understanding systems of laws.

Every anthropologist knows that early groups of humans are
postulated to have been hunters of large game. Their predeces-
sors almost certainly lived in groups for the reason that, like
probably all modern group-living nonhuman primates, they
were the hunted rather than the hunters. By all indications we
are the only primate that became to some significant extent a

group-hunter—the only group-living primate who, at least for a time, escaped having our social organization essentially determined by large predators. In this light it may not be so startling that dog and wolf packs and lion prides are social groups with which we empathize to a very great degree—social groups that fascinate behaviorists because of parallels and complexities that are not clearly established elsewhere outside the human species. Our human brand of sociality thus appears to be approached from two different directions—by various other primates because they are our closest relatives, and by canines and cats because they most nearly do, socially, what we did for some long time.

But the organization and maintenance of recent and large human social groups cannot be explained by a group-hunting hypothesis (Alexander, 1971, 1974). The reason is obvious: the upper size of a group in which each individual gained because of the group's ability to bring down large game would be rather small. Indeed, according to such a hypothesis, as weapons and cooperative strategies improved, group sizes would have gone down, owing to the automatic expenses of group-living. Instead they went up—right up to nations of hundreds of millions.

Human nations of millions of individuals, each potentially reproductive, appear to be unique in the history of the earth. There is no parallel, as often supposed in the past, with the social insects, in which one or a few females do all of the reproducing and the rest are closely related sterile workers and soldiers. Chimpanzees, baboons, and macaques are probably our closest counterparts in this regard, and their social organization probably was strongly affected by predators of other species (one of the most important of which may well have been our own ancestors).

What, then, did cause human groups to keep right on growing? If we hold to the arguments about group-living described in the first chapter, what forces could possibly account for the rise of what anthropologists have called the "nation-state"? The uniqueness of human group sizes, as well as the uniqueness of humans, suggests that unique, truly remarkable causes may be involved.

One possibility is that the early benefits of group-living (such as group-hunting, cooperation in farming or fishing, and a host of others) were so powerful that they produced humans with such strong tendencies to group that the huge modern nations of today developed as more or less incidental effects. This argument is valid only if humans are considerably less flexible in their behavior than I would like to allow. It says, in effect, that we are captives of our genetic history, and are so compulsive about group-living that we pursue the habit relentlessly despite deleterious effects on ourselves and our children—despite its hindering effects on the reproduction of ourselves and our close relatives. Perhaps such an argument does not seem too remote to those who have regarded reproduction as a triviality in human history, or to those who do not recognize the degree of opportunistic flexibility that typifies the human organism. Any such argument, however, seems entirely impotent to me. Moreover, should this be the real reason for human sociality, alternative hypotheses should not easily apply.

But there is an alternative hypothesis, one recently proposed several times by different authors, and one that seems reasonable, appropriately unique, and clearly relevant to all efforts to understand, govern, and perpetuate ourselves (Keith, 1949; Alexander and Tinkle, 1968; Bigelow, 1969; Carneiro, 1961; E. O. Wilson, 1973a, 1975; Durham, 1976b). I will call it the "Balance-of-Power Hypothesis." This hypothesis contends that at some early point in our history the actual function of human groups—their significance for their individual members—was protection from the predatory effects of other human groups. The premise is that the necessary and sufficient forces to explain the maintenance of every kind and size of human group above the nuclear family, extant today and throughout all but the earliest portions of human history, were (a) war, or intergroup competition and aggression, and (b) the maintenance of balances of power between such groups. I emphasize that this is hypothesis not conclusion, and I state it in this simple radical form to make it maximally vulnerable to falsification. I am not implying that no other forces influence group sizes and structures but that balances of power provide the basic sizes and kinds of groups upon which secondary forces like resource distribution, population densities, agricultural and technological developments, and effects of diseases exert their influences. And I am suggesting that all other adaptations associated with group-living, such as cooperation in agriculture, fishing, or industry, are secondary—that is, that they are *responses* to group-living and neither its primary causes nor sufficient to maintain it, at least in a world not so densely crowded with humans that there is essentially no way to live alone. Social scientists may regard this kind of effort to locate a single primary adaptive significance as unsupportably reductionistic and oversimplifying, but in biology it is the usual approach to questions of evolutionary function. Even if it turns out to be too simple to account for human groupings I believe it is the right way to start, and more likely than any other approach to give us ultimately the correct answers.

This argument would divide early human history into three periods of sociality, roughly as follows:

1. Small, polygynous, probably multi-male bands that stayed together for protection against large predators (polygyny does not necessarily imply the maintenance of harems, but simply that fewer males than females were contributing genetically— Trivers, 1972).

2. Small, polygynous, multi-male bands that stayed together both for protection against large predators (probably through aggressive defense) and in order to bring down large game (perhaps, at some times, entirely because of one or the other of these reasons).

3. Increasingly large polygynous, multi-male bands that stayed together largely or entirely because of the threat of other, similar, nearby groups of humans.

I suggest that expressions of human social organization today are the legacies of this sequence, with the relative importance of each stage in understanding sociality in modern humans dependent upon the duration of the stage, the intensity of selection during it, and where it occurs in the sequence. I also suppose that we have been in the third stage so long that the influences of the first two stages are relatively minor. The latter assumption departs dramatically from arguments of other authors, but that departure is not critical to the arguments which follow. Thus, Eaton (1978) suggests that early humans may have spent a very long time during which their social behavior was largely structured by both defense against large predators and competition with them. It seems to me that the idea that the complex behaviors required for such activities could have "primed" or preadapted humans for their later evolution in hostile intraspecific groups, whose chief threats were one another, is entirely compatible with the argument I advanced first in 1971 and have reviewed here.

To relate the scenario I have just constructed to the thinking of archaeologists and anthropologists on the problem of the rise of nations I would call attention to Flannery's (1972) and Service's (1975) discussions of the evolution of civilizations, to Carneiro's (1961) paper, to Webster's (1975) critique of Carneiro's argument, and to Wright (1977). Flannery describes the modern range of human societies from small hunting-gathering bands of less than two hundred individual affiliates through tribes and chiefdoms to huge industrial nations. He notes that all large nations had to pass through at least some of the smaller stages to reach their present condition, and shows that archaeological evidence regarding the earliest known dates for the three classes of societies larger than band-societies suggests the appropriate chronology from small groups to large (figure 12). Then he asks what "prime mover" could account for the trend toward larger, more complex states and nations?

After reviewing many extrinsic possibilities and finding each either unnecessary or insufficient, Flannery seems to follow an approach frequently resorted to by biologists and social scientists alike in this kind of situation; he seems to seek the reasons for the nature of society in its internal workings. This approach leads to hypotheses like the one discarded earlier here—hypotheses of orthogenesis or genetic, physiological, and social "constraints" or "inertia." In biology it leads to what are termed arguments from physiological limitations—tendencies to explain each attribute as the maximum that could be achieved in a certain direction despite continued favoring of directional change. In effect it requires that one explain ultimate causes by proximate causes rather than vice versa, and it is at best a vulnerable argument.

To invoke proximate limitations to explain extant phenomena of life is in effect to deny the power of the evolutionary process to produce some perceived or imagined effect. Sometimes there are valid arguments from adaptation for such explanations. An example is the argument that no more than two sexes exist because the presence of three or more would automatically cause an ecologically inferior sex to become less valuable as it became rare, and eventually to disappear; with two sexes individuals of the rare sex automatically become more valuable (Power, 1976). But to deny adaptive significance because of supposed constraints on natural selection is perilously close to asserting either that marvels like humans and honeybees are impossible, or that they are entirely predictable. Moreover, the rapid directional changes induced by human selection, especially upon domestic animals and plants, and the diversity of effects achieved within species by different directions of selection in only a few generations, tend to deny any long-term significance of genetic constraints and attest to the potency of selection (Fisher, 1930; Hamilton, 1967; Trivers and Willard, 1973; Trivers and Hare, 1976; Alexander and Sherman, 1977; Alexander et al., 1979).

How, though, does Flannery dismiss the hypothesis suggested here, that of intergroup competition? He notes that intergroup aggression has evidently been continual throughout history in many parts of the world where large nations have never evolved. Like Webster (1975) and others (Service, 1971; Carneiro, 1975), he concludes from this that while war may be necessary to explain the origin of the state, it is not sufficient. As Wright (1977) put it: "Warfare as a motivation for cooperation and an eliminator or subjugator of less effective organizations is universal in human development and cannot by itself explain state emergence" (p. 380).

But these authors do not explicitly consider the question of balance of power. Balances of power depend to some extent on physiographic and other extrinsic environmental circumstances, and they may as well exist between tiny New Guinea tribes as between nuclear powers (Carneiro, 1975). Moreover, aspects of intergroup conflict among such people, which are commonly referred to as ceremonial or ritualistic, may actually reflect the importance of balances of power; examples are elaborate bluffing and the intensity of concern with avenging each death. Balances of power are also significant within groups, continually denying to individuals and subgroups the possibility of initiating individualistic reproductive strategies or of fragmenting the larger group by secession or fission.

If, for whatever reasons, recurring imbalances are not possible through one-sided expansions of first one group, then the other, or through alternations of superiority in weapons or other regards, then the balance-of-power races leading to large nations may never appear. One test of a balance-of-power hypothesis would involve checking to see if physiographic or other barriers reduce the effectiveness of coalitions or the likelihood of unity across areas of increasing size, preserving the balance at small group sizes (Carneiro's "environmental circumscription"). Another test would be to see if empires have tended to develop in pairs or groups, or centrally nested inside multiple smaller competitors, and to disintegrate when they lacked suitable adversaries (Carneiro's "social conscription"). Even a very general knowledge of history suggests that these things have been true. Carneiro's hypothesis, and the analyses of Flannery and others, seem to me to put us on the brink of modifying to acceptability a hypothesis of just the sort discussed here (and in somewhat different forms by others [Keith, 1949; Carneiro, 1961; Alexander and Tinkle, 1968; Bigelow, 1969; Alexander, 1971, 1974; E. O. Wilson, 1973a, 1975; Durham, 1976a,b]).

Some authors argue that there is "absolutely no evidence" that intergroup competition and aggression have played a central role in human evolution (e.g., Montagu, 1976, and personal communication). This stance puzzled me until I realized that two kinds of evidence bear on this question and authors such as Montagu are depending chiefly on one kind while others including myself rely chiefly on the second. One kind is physical evidence of aggression, including fossils. Little or none of this evidence is unequivocal: spear points, arrowheads, and stone axes all have been called "tools" or "weapons," depending on one's bias, and they could have been either or both; skulls could have been crushed by predators or damaged after death; evidence of cannibalism could have been interpreted differently if it came from ceremonial affairs within groups rather than from wars; etc.

On the other hand, even continuous intergroup hostility and aggression do not necessarily leave a record for archaeologists to trace. If there were no written records, what evidence would there be to tell us what happened to the Tasmanians and the Tierra del Fuegians? Without written records could we have been unequivocal thousands of years later about what the invading Europeans did to the Native Americans on both continents of the New World? Consider the most monstrous cases of genocide in recorded history: can we even be sure, again without written words, that what happened in the twentieth century at Buchenwald and Auschwitz, and in Nigeria and Cambodia, will be properly interpreted, say, a million years from now? Yet more people may have been killed in these places than existed in all of the time before recorded history. Such questions, it seems to me, cast doubt on the interpretation that equivocal evidence of human aggression, not to say the milder yet potentially continual and crucial forms of intergroup competition, must automatically be discarded.

The second kind of evidence comes from interpreting recent history and the behavior of modern humans, and then asking about the legitimacy of extrapolating backward in time, both to postulate what happened and to interpret the otherwise equivocal evidence from archaeology and paleontology. We know that intergroup competition and aggression have been continuous across nearly the whole face of the earth throughout recorded history. We know that cooperativeness on the grandest scale, and the greatest of all of the alliances of history, were in response to upsets in balances of power and the aggression of one nation against another. We know that competition is continuous among the various kinds of political groups, large and small, that exist across the whole earth. We know that atomic fission, space travel, and probably most of the remarkable modern advances in science and technology occurred or were accelerated as a consequence of intergroup competition or outright war.

These facts do not imply that humans are bound to war forever, that we are instinctively destined to keep on killing one another. They do not mean that our aggressiveness is not learned and not changeable or even not removable, nor that war or fighting or desperate competition is good. These are red herrings raised by those either compassionately or politically motivated to deny our history of aggression and killing, or wishing to counteract the declarations of others who support "innate-aggressive" arguments either naively or for their own pernicious purposes.

Along with our obvious abilities and tendencies to be either aggressive and deadly or cooperative and altruistic as the occasion demands, the facts of history cited above are evidence, in my opinion, that when extrapolating into earlier times known only through fossil evidence, we cannot properly assume that humans were socially different then. I do not believe the burden of proof is upon those of us who see humans as evolving while behaving more or less as they do today. It is the other way around, and mere absence of unequivocal paleontological and archaeological evidence of aggressiveness and competition is not sufficient to suggest that these things did not occur; indeed, it is proper to demand instead that the evidence of no aggression be unequivocal, and it is not. In my opinion, even equivocal evidence of aggression, when coupled with the events of recorded history, supports the argument that competition and aggression have indeed been powerful factors throughout human evolution. Beyond all other reasons, to suggest that these kinds of interactions have not been important is to argue that our ability to learn how to conduct and use aggression and competition in almost unbelievably clever and diverse ways is only an evolutionary accident. In a way it argues the same for our ability to cooperate, since we seem to do that best and most massively in intergroup competition. I for one believe there is not an iota of evidence to support the idea that aggression and competition have not been central in human evolution.

Those who reject arguments like those given here also point to what they interpret as relatively nonaggressive behavior on the parts of the hunting and gathering societies that remain today in a few places like the Australian and African deserts, and the Arctic. They argue that for 99 percent of their history our ancestors lived as these people do. But such people survive today only in marginal impoverished habitats that support only the lowest of all densities of human population and also represent physical extremes that by themselves require cooperation among families for mere survival; moreover, hunter-gatherers survive today only because even the most advanced technological societies have found no way to use their homelands that would make it profitable to overrun or seize them by force. In other words, they are restricted by aggression, or its threat, to the localities where they exist now.

Nevertheless, social competition, aggression, murder, and intergroup conflict are not by any means unknown among hunter-gatherers (Ember, 1978). I suggest that the ordinary interpretation may be precisely wrong, and that rather than all humans having spent 99 percent of their existence living as Eskimos and Bushmen do today, the ancestors of Eskimos and Bushmen more likely spent most of their existence in richer habitats where higher densities of population, more complex social structure, and less harsh physical environments led to both more complex and extensive cooperativeness and more complex and extensive social competition than now exists. This is a virtual certainty for Eskimos and Australian Bushmen, and I know of no reason to doubt it for Africans. That a human society survived by hunting and gathering, moreover, is no reason by itself for arguing that it was either aggressively competitive or not, and that such societies are sometimes nonaggressive in the ecological outposts of the modern world is no reason for supposing that this was always so in the more resource-rich habitats of the world where humans also lived and developed all of the complex civilizations of history.

The arguments made here are supported by those of Ember (1978), who refers to the anthropological view that hunter-gatherers are peaceful and noncompetitive as a "myth": she presents evidence that 90 percent of the hunter-gatherer societies engage in intergroup aggression almost as frequently as the rest of the societies of the world. Ember's data suggest that aggressive behavior in hunter-gatherers, as in other kinds of societies, varies as the need for it, or the possibilities of gaining by it, vary.

The criticism is sometimes made that Flannery's (1972) kind of reconstruction assumes that a particular modern ethnographic example is an exact replicate of its archaeological (and extinct) counterpart, or even, in the extreme, that the ethnographic and archaeological examples are implied to give rise to one another, always progressing from simple to complex. This is the so-called "unilineal evolution" model of societal change, which has been likened to the notion that history is the development of an embryo. But Flannery's analysis is not a defense of unilineal evolution, nor is the reconstruction proposed here. To level such a criticism at every effort to locate "prime movers" implies a basic misunderstanding of comparative method. Comparative method, in modern biology, sociology, or anthropology, assumes: (1) that sequences of change have occurred (genetic evolution and cultural change); (2) that parallel sequences of change occur in different places, at different times, and in different lines at the same times and places; (3) that some (but not all) of the attributes of different stages (but not the actual cases or even, necessarily, the actual sequences) will be represented in both extant and extinct forms; and (4) that appropriate comparisons of such attributes can yield information about the sequences of change and their causes. These assumptions allow interpretations of the past by studying the present, or vice versa, and comparative method, explicitly in the sense described here, represents a main source of evidence for evolutionary biology, archaeology, sociology, and cultural anthropology (for a parallel opinion from sociology, see Marsh, 1967). As Hoebel (1954) remarks:

> . . . in its own particular history *a* society does not have to go through all the successive steps of the technological sequence. Borrowing may make great leaps possible. Eskimos are today serviced by airplanes and steamships. They moved from simple hunting savagery into a mechanical civilization within the span of a hundred years. But—in the evolution of culture, collecting and hunting came first; they endured throughout the Old Stone Age. . . . The first domestication of plants and animals did not take place until Neolithic times. . . . And the Machine Age got its start hardly more than three centuries back [p. 292].
>
> A society with a hunting culture *is* more primitive and less evolved than one with hoe-culture or simple pastoralism. These in turn *are* more primitive than one with agriculture or higher pastoralism; and these in turn *are* more primitive than one with industrialization [p. 292].

My only reservation in regard to this statement—and I believe it is a significant one—is that we must keep in mind the possibility that some kinds of seemingly "more primitive" societies may sometimes be derived from "less primitive" societies. Since, for example, as Hoebel also notes, "Grasslands and semideserts do not lend themselves to primitive gardening" (p. 292), it is at least possible that some hunter-gatherers are descended from primitive gardeners, etc.

Across the past several decades failures in the social sciences to locate broad explanatory generalizations have led to the supposition that they do not exist, that it is more appropriate to seek or rely upon multiple causes than to accept singular ones

—even, it appears, if one of the latter should seem sufficient! Perhaps it seems vile and degrading that there may be singular explanations for complex phenomena of human behavior, but not nearly so much so as to deny them on that basis alone. The new evolutionary arguments about group-living summarized above, for example, not only cast doubt on the older "group function" explanations but also imply that a singular explanation is both possible and likely.

Let me review the steps by which I arrived at the hypothesis that the rise of the nation-state depended on intergroup competition and aggression, and the maintenance of balances of power with increasing sizes of human groups. First, Williams' (1966) convincing argument that selection usually is effective only at individual or genic levels forced a search for reasons for group-living that would offset its automatic costs to individuals. The available reasons have proved to be small in number, and only one, predator protection, appears applicable to large groups of organisms, including humans. For humans a principal "predator" is clearly other groups of humans, and it appears that no other species or set of species could possibly fulfill the function of forcing the ever-larger groups that have developed during human history. Carneiro (1961) and Flannery (1972) essentially eliminated as "prime movers" all of the other forces previously proposed to explain the rise of nations, and I think their arguments are reasonable. Flannery (1972) and Webster (1975) also eliminated intergroup competition as a prime or singular force, and they sought causes of the rise of nations within societal structure. This last procedure I see as unsatisfactory because it involves explaining ultimate factors by proximate mechanisms. Flannery's rejection of intergroup aggression as necessary but not sufficient is inadequate because he did not specifically consider intergroup aggression in terms of the maintenance of balances of power. His elimination of other factors may or may not be satisfactory; realizing, however, that automatic expenses to individuals accompany group-living, expenses that are generally exacerbated as group sizes increase, I find that none of the supposed causes for the rise of nations except balances of power seems even remotely appropriate. Nor do they serve any better when grouped and regarded as multiply contributory.

Some social scientists seem to believe that to locate function at group rather than individual levels implies less competitiveness and strife during human history. This is not necessarily true, and the above arguments indicate that in some senses the implication is just the opposite. After all, closely knit groups can be much more destructive than independently acting individuals, and group behavior when function is largely at the individual rather than the group level, whatever else it means, also implies majority rule, suppression of bigotry, and preservation of individual rights. Evidence of intergroup selection in human evolution with a concomitant emphasis on intragroup altruism is not necessarily cause for optimism about the future since it presages certain difficulties in discovering environments in which altruism within groups can be induced without the simultaneous fostering of hostilities between groups. In today's world the latter is certainly more generally dangerous than the former, and probably the only real threat to civilization.

Here I insert a caution: what I am saying throughout this book may be right or wrong as an interpretation of human history. In any case it does not imply a deterministic future. On the contrary, I have argued that the individuals and groups least bound by history are those who best understand it. Hence, the question is only whether or not the view of history suggested here is correct. I reject any notion that what I am saying leads to an acceptance or promotion of a social Darwinist future, or is pertinent to the social designs of any particular ideology. Those who would abandon my arguments about history on such grounds are much more deterministic in their view of human behavior than I am.

Group-Living and Rules

Besides serving as a means of channelizing collective action and social control, which justifies it to the governed, . . . [government] offers to those in authority

> *an opportunity to use their power for*
> *selfish aggrandizement. To the barbaric*
> *chieftain, the feudal lord, and the munici-*
> *pal boss, alike, accrue special privilege*
> *and pelf.*
> —*George Peter Murdock, 1949, p. 84*

The arguments developed so far lead us to hypothesize that rules in some fashion represent the wishes of individuals and relate to reproductive competition among individuals within groups, with the additional constraint that individual reproductive success within a group depends to some extent on the success or maintenance of the group as a whole. In other words, we might hypothesize that individuals behaving consistently in respect to the long history of human evolution should work to preserve their group and keep it healthy while simultaneously striving so far as possible (of course, not necessarily consciously) to populate it with descendants and other genetic relatives as closely related to themselves as possible.

In support of this view, we may note that efforts to cause changes in the behavior of populations only work well or for long when the individuals in the population regard the changes as personally advantageous: it has to be to the *individual's* advantage to reduce family size, conserve fuel, or treat his neighbor right; or it has to be to his disadvantage not to do so. Cooperative subgroups, such as corporations, are not likely to follow courses matching the interests of the whole group, as in pollution, resource depletion, or profiteering, unless (a) the penalties imposed by the whole group are sufficient to eliminate the profit in selfish behavior, or (b) a threat external to the entire group makes it temporarily profitable for the subgroup to direct efforts primarily to sustaining the larger group (not the least reason for which is the "public relations" effect from altruism or heroism in such times) (Fisher, 1930).

It would seem to be the purpose of laws to impose such penalties, to meet such threats, and to be regarded by individuals as having these functions. Whenever a group is plainly threatened by external forces, so that the individual and his kin are also threatened, then individual members tend, with the least encouragement, to obey the laws and work for the common good. Indeed, under such circumstances even huge nations form alliances with one another.

WHAT IS JUSTICE?

To relate these musings by a biologist to the thoughts of legal philosophers and sociologists we can turn briefly to perhaps the most widely asked question about societal laws, the very prominence of which supports the individualistic interpretation of history being defended here. The question is, "What is justice?" I will start with two excerpts from an essay by the philosopher Hans Kelsen, which opens his book *What Is Justice* (1957). Kelsen notes:

> No other question has been discussed so passionately; no other question has caused so much precious blood and so many bitter tears to be shed; no other question has been the object of so much intensive thinking by the most illustrious thinkers from Plato to Kant; and yet, this question is today as unanswered as it ever was. It seems that it is one of those questions to which the resigned wisdom applies that man cannot find a definitive answer, but can only try to improve the question. [P. 1]

In a discussion relating justice and "social happiness," Kelsen focuses on the crucial point, conflicts of interest:

> It is obvious that there can be no "just" order, that is, one affording happiness to everyone, as long as one defines the concept of happiness in its original, narrow sense of individual happiness, meaning by a man's happiness, what he himself considers it to be. For it is then inevitable that the happiness of one individual will, at some time, be directly in conflict with that of another. . . .
> Where there is no conflict of interests, there is no need for justice. A conflict of interests exists when one interest can be satisfied only at the expense of the other; or, what amounts to the

same, when there is a conflict between two values, and when it is not possible to realize both at the same time; when the one can be realized only if the other is neglected; when it is necessary to prefer the realization of the one to that of the other; to decide which one is more important, or in other terms, to decide which is the higher value, and finally: which is the highest value. [Pp. 2–4]

Kelsen concludes that justice must be relative and incomplete, and can only be regarded as ideal or absolute if it is accepted by everyone as having been determined by an ideal or absolute being, such as God—a view expressed by a long succession of philosophers.

Justice is necessarily incomplete, and laws are fluid, then, because people strive, and their strivings conflict. To understand the nature and extent of the conflicts, hence, sociality and the sociology of law, it would seem useful to know what they are striving for. This too is an old question. Kelsen, and the Declaration of Independence, say they are striving for happiness, and this is no doubt true. Happiness, though, as the different versions of a familiar adage tell us, is many things. It is eating and sex and parenthood and warmth and touching and ownership and giving and receiving and loving and being loved; it is the cessation of pain and the onset of pleasure; it is finding a way to win; it is having a magnificent idea or a grandchild. In some sense these are all biological things. There is probably nothing on the list that does not bring a happiness equivalent to at least one nonhuman organism as well as humans.

All that modern biology is adding to this set of questions and arguments is the suggestion that humans have evolved to strive to reproduce, and to reproduce maximally—indeed, to *out*reproduce others. Happiness, then, is an end for the individual only in the sense that it is achieved by acts leading to reproduction. Happiness is a means to reproduction (that this is strictly true only in historical terms and that happiness can obviously be diverted from the goal of reproduction can be ignored for the moment).

In other words, the striving of organisms can be generalized with confidence, and it is not hedonistic at all but reproductive; in historical terms hedonism is itself reproductive, and when it is not we expect it eventually to be abandoned.

I believe that these thoughts give us a way of understanding why human striving is incompatible with the concept of ideal, pure, or complete justice. First, humans strive as individuals or subgroups rather than as united wholes. Second, there is no automatic finiteness to their striving because success can only be measured in relation to the success of others. It follows that their separate strivings conflict, and sometimes involve direct thwarting of one another's efforts. Finally, they are continually altering their strivings to increase their success as circumstances change and thereby introducing additional changes.

The differences of interest that legal philosophers discuss are thus based on differences of reproductive interest, and ultimately on genetic differences, and they are not likely to be resolved, in any absolute sense, by allowing certain amounts of reward, payment, or return for given contributions.

We will have to examine, then, several items, and the first is degrees of genetic relatedness or overlap among interacting individuals. We can even suggest that it is no accident that Kelsen's examples of the complexities of justice are: (a) two men in love with the same woman, (b) King Solomon's threat to divide a disputed child between the two women who claimed it, with the intent of giving it to the woman who loved it too much to allow it to be hurt, and (c) two men in competition for the same prestigious job.

Since most people today live in nation-states, we must also be interested in the nature of such societies and the possibility of generalizing the basis for the systems of law by which nation-states manage to function. Hoebel (1954, p. 275) accepts law as the aspect of our culture "which employs the force of organized society to regulate individual and group conduct and to prevent, redress or punish deviations from prescribed social norms." He describes four functions of law, three of which are the direct regulation of behavior and the fourth "to redefine relations between individuals and groups as the conditions of life change." Stein and Shand in their book, *Legal Values in Western Society* (1974), argue that order is "the primary value with which law is associated." But they answer the question "What is law for?" by saying that the "three basic values of the legal system" are "order, justice, and personal freedom." From the arguments I have just made, and those that follow, I suggest that law is "for" but one thing: the preservation of order; and that justice and personal freedom, to whatever extent they are sought or approached, are also for the purpose of maintaining order. Order is valuable to everyone if extrinsic threats to the group are sufficiently severe, and the group is of no value if there are no such threats. In times of little or no extrinsic threat, laws are most valuable to those who lopsidedly control resources. These people generally include the wealthy and secure (versus the poor and insecure), parents (versus individual offspring), and older people (versus younger people). To be a revolutionary (that is, to be willing to destroy order) individuals must either (a) perceive themselves to be in a very bad position within society, (b) suppose that no significant threats to the group exist at the moment (so that internal dissension would not lead to worse troubles for themselves as a result of outside forces), or (c) have support from outside the group that seems to them to promise a better situation as a result of destruction of the existing order, and perhaps even destruction of their group as constituted.

And so we are returned to the biologist's view of organisms' lives being divisible into the activities of resource-gathering and resource-distribution. I think this view yields a significant new perspective about the basis for innumerable everyday phenomena like inheritance laws, changes in occupation at mid-life, racism, and reasons why racism's effects fall more heavily upon males of the minority group than females.

Justice, Happiness, and Keeping Up with the Joneses

The legal philosopher John Rawls (1971) accepted and developed the idea that justice correlates with happiness, and suggested that happiness may be identified as follows: "A person is happy when he is in the way of a successful execution (more or less) of a rational plan of life drawn up under (more or less) favorable conditions, and he is reasonably confident that his intentions can be carried through. . . . adding the rider that if he is mistaken or deluded, then by contingency and coincidence nothing transpires to disabuse him of his misconceptions" (pp. 548–49).

But Rawls failed to consider fully how individuals decide upon particular courses of action, and why there is any likelihood at all of selecting a plan of life that is *not* likely to be carried through, particularly in an affluent society where scarcely anyone is actually in danger of starving, freezing, or otherwise dying prematurely because of inability to secure necessary resources. In other words, he failed to explain why people strive, and what he left out seems to be the crux of the problem, and the source of the conflicts of interest that lead to ethical questions. I think we can be certain that, even in affluent societies—and, I would venture, *especially* in some such societies —there will be much evidence of unhappiness. Why should this be so?

From a historical perspective, success is only measurable in relative terms. We set our goals and determine our plans on the basis of what we observe others about us achieving. Such goals can be irrational or inaccessible, and thus lead to unhappiness, when individuals (a) strive from inferior resource bases, and fail to consider the obstacles placed in their way (because of race, sex, physical or mental handicaps, or other bases for discrimination), (b) fail to take into account trends in society which change the structure of opportunities, or (c) fail to consider the extent to which others' achievements have required inordinate power, influence, chicanery, or injustice against others (and the attendant risks). I think we can predict that unhappiness as a consequence of irrational personal goals is likely to be most prevalent in societies that are hierarchically structured, in which lofty goals may be developed from observations of others' success, and in which severe inequality of opportunity renders these goals inaccessible for what are logically interpreted as unjust reasons.

In natural selection the likelihood of a genetic element persisting depends entirely on its rate of change in frequency in relation to its alternatives; changes in absolute numbers are irrelevant. Among the attributes of living creatures, whatever can be shown to have resulted from the action of natural selection may be expected to bear this same relationship to its alternatives. Thus, we should not be surprised to discover that the behavioral striving of individual humans during history has been explicitly formed in terms of relative success in reproductive competition, that justice is necessarily incomplete, that happiness is not easily made universal, and that ethical questions continue to plague us, and can even become more severe when everything else seems to be going well.

> Since status is in such hopelessly short supply, life appears to be a zero-sum game, in which one man's loss is another's gain.
> —Silberman, 1978, p. 114

REPRODUCTIVE COMPETITION AND LAW-BREAKING

I have argued that *the function of laws is to regulate and render finite the reproductive strivings of individuals and subgroups within societies, in the interest of preserving unity in the larger group* (all of "society" or the nation-state). Presumably, unity in the larger group feeds back beneficial effects to those segments or units that propose, maintain, adjust, and enforce the laws. Partly because of continual shifts of interest, changing coalitions, and power adjustments, group unity is not likely always to feed back evenly, or in such fashion as to cause all individuals to benefit equally from that unity; under such conditions "federal" government becomes possible.

The general theory that laws function to place limits on reproductive striving leads to several predictions. First, laws should be seen as constructed so as to regulate competitive striving, and the severity of punishment is expected to reflect the severity of deleterious effects on the reproduction of others. Capital punishment is generally correlated with such transgressions as murder, which destroys the victim's ability to reproduce any further; treason, which threatens to lower the fitnesses of everyone else in one's society; or rape, which may directly interfere with a man's chances of reproducing via his spouse, sister, daughter, or other female relatives. Rape laws are particularly revealing. Because rape as such does not lead to anyone's death, rape does not at first seem an appropriate transgression for the imposition of capital punishment.

If we were to select a category of striving that is centrally important, restricted to a definite part of the human life-span, and more intense in one sex than the other, then we should be able to plot changes in the intensity of that striving—say, by age—against changes in the likelihood or rate of lawbreaking. An appropriate topic is sexual competition, or competition for mates, including all of the various activities involved in increasing one's desirability as a mate, hence, one's ability to select among a wider array of potential mates. Sexual competition is demonstrably more intense among males than among females; and one can easily show from the accumulated differences between modern males and females powerful evidence that this has consistently been the case during human history, and that as a general consequence the entire life strategy of males is a higher-risk, higher-stakes adventure than that of females (Alexander et al., 1979). This finding leads to the prediction that lawbreaking will occur more frequently among males, which of course is already well known (Sutherland and Cressey, 1966). It also seems to predict that laws are chiefly made by men (as opposed to women) to control men (as opposed to women).

That laws are made by men to control men is suggested over and over again in their structure and application. As perhaps the prime example, it has become painfully clear, because of attention recently brought to bear by women's groups on the application of laws against rape, that the female victims are treated like pawns by the collections of (mostly) males who enforce the laws, judge and punish the offenders, and are indirectly wronged because of affinal or kin relationship to the victim. One might say that rape victims appear to be treated as if rape laws were designed to protect them only in the sense that rape wrongs the males to whom they "belong" or might have belonged and reduces their value or attractiveness to those males; and in the sense that when rapists are free to act, the interests of all *men* are in jeopardy. It appears that the female victims of rape may be only incidental to the development and application of rape laws. Under current circumstances the most pathetic of all rape victims is probably the female in one of the following situations: (1) she is without a male who has a proprietary interest in her welfare—such as a father, brother, husband, or sweetheart; (2) her male defenders are already somewhat tentative in their allegiance to her; (3) she was raped by such a person or in such a circumstance as to pose little threat to other males in society; or (4) she was raped in a fashion or circumstance that reflects particularly severely on her future desirability as a mate. It is relevant that in many states a man cannot be accused of raping his wife, and, until very recently, a man could kill his spouse or her lover without being accused of murder while a woman could not do the same.

Lawbreaking is also expected to be concentrated at those periods in life, or those ages, when competitive striving is most intense or most crucial. Competition for mates is greatest just before the usual age of marriage, and the extent to which an individual is able to begin effectively to climb the ladder of affluence may also be determined at about the same ages. Lawbreaking is strongly concentrated during ages seventeen to twenty-two in technological nations (Mulvihill and Tumin, 1969). This is also the age span during which Yanomamö Indians (Chagnon, 1968), and probably men in most societies, suffer their highest mortality, mostly from intergroup aggression, and also the age of highest likelihood of military induction. As predicted, this age level immediately precedes that at which marriage is most frequent.

Lawbreaking is expected to be higher in individuals or groups most inhibited from climbing the ladder of affluence or using the system legally to accumulate resources. Moreover, lawbreaking should be even more heavily concentrated in males who are more or less publicly identified as likely to have such difficulties. Thus:

1. Lawbreaking should be, and is, higher in minority-group males than in majority-group males (Ferracuti and Dinitz, 1974; Fleisher, 1966; Silberman, 1978)

2. Lawbreaking should be, and is, higher in publicly (e.g., physically) identifiable minority-group males than in those not publicly identifiable (Himelhoch, 1972; Sutherland and Cressey, 1966; Mulvihill and Tumin, 1969; Silberman, 1978)

3. In the absence of publicly identifiable minority groups, lawbreaking should be highest in young males whose families give them least assistance in climbing the ladder of affluence. The most dramatic correlation found by Ferracuti and Dinitz (1974) with delinquency in a racially homogeneous Puerto Rican situation was with *lowering* social status of the boy's family (see also Himelhoch, 1972; Cortes and Gatti, 1972; Clark and Wenninger, 1962; Fleisher, 1966). (See also the reference, elsewhere in this book, to lawbreaking among the children of immigrant workers in Germany.)

Lawbreaking is also higher in families lacking one parent—especially the father—in families that are less religious, and in families with less control over their children and who give their children less assistance, encouragement, and attention, but who punish them more often (Ferracuti and Dinitz, 1974; Himelhoch, 1972; Cortes and Gatti, 1972; Clark and Wenninger, 1962; Fleisher, 1966). These findings are not inconsistent with the above arguments, for absence of a parent and "extremes of laxity and harshness" in the socialization of children all suggest what Glueck and Glueck (1950) term "the greater inadequacy of the parents of the delinquents" (p. 280).

Finally, we may consider alternative strategies of reproductive competition among males, their distribution, and their consequences for patterns of lawlessness. I assume that freedom of opportunity to "climb the ladder of affluence" is a crucial aspect of sexual competition which, when available, will supplant all other aspects—that is, that males who are either affluent or have a very high likelihood of becoming affluent are among the most desirable mates. Such males are unlikely to be lawbreakers, except in the context of using the system to further their affluence illegally, such as by income tax fraud or misuse of power associated with affluence.

In contrast, one expects alternative strategies, such as behavior that can be considered under the general label of "machismo," or flash and braggadocio, to be concentrated in individuals or groups whose likelihood of climbing the ladder of affluence ("using the system" effectively) is lowest, and especially when this low likelihood is publicly projected by inescapable identification with a disadvantaged group, such as a minority against which the system discriminates. "Macho" strategies of sexual competition are at once declarations of desirable qualities other than affluence, denials of the value of affluence as usually measured, rejection of the system, and declarations of a degree of disdain and independence with regard to the rules of the system. Sexual competition by "macho" behavior is almost by definition a declaration of lawlessness, or willingness to break the law.

Cohen (1955) documents "the disproportionate concentration of delinquency among the less prosperous, powerful and respected," notes that the "delinquent subculture" is very predominantly male, regards "the family as a unit in the social class system," and concludes that "the *kinds* of achievement or exploit which are valued differ. . . . [and] they are harnessed, disciplined, systematized and put in the service of long-run goals in the middle-class culture; in the delinquent subculture, they are affiliated with and transformed by impulsivity, short-run hedonism, violence and predation" (p. 139).

Silberman (1978), in an extensive and excellent review of this topic, writes as follows:

Street criminals tend to be young and poor, with a large proportion coming from minority-group backgrounds. In 1976 . . .
nearly 75 percent of those arrested for the seven serious crimes
included in the FBI's Crime Index were under twenty-five years
of age, and more than 40 percent were seventeen or younger.
. . . One-third of those arrested were black, and Puerto Ricans and
Mexican-Americans made up a significant proportion of arrestees
classified as white.

Street criminals are also predominantly male: in 1976, women
comprised only 7 percent of those arrested for robbery, and 5
percent of those arrested for burglary [p. 49]. . . .

Being "cool" is important to lower-class males, for it is synonymous with courage. Action, in whatever form, provides a chance
to demonstrate their ability to face a challenge and overcome it,
and hence to offset the impotence they normally feel. At the very
least, the excitement that action generates provides evidence to
the individual that he is alive, and that significant others know
that he exists [pp. 112–13]. (See also Miller, 1958; Goode, 1966;
Merton, 1968; McCord and McCord, 1969; Staples, 1978.)

In summary, the predictions appear to be met that "macho"
behavior and lawbreaking are concentrated in men who are
young, lack wealthy, influential, successful, or powerful relatives, and are recognizable as members of minority or other
disadvantaged groups. It is an obvious corollary that somehow
to equalize the possibilities for individual men of all classes and
origins to "climb the ladder of affluence"—at least in terms of
personal capabilities (and how they are seen by their posses-
sors)—would provide the most reliable if not the easiest way of
reducing the problem of criminal behavior. Perhaps not surprisingly, this conclusion is consistent with those arrived at from
entirely different approaches, and after very extensive exami-
nations of correlates of crime and delinquency (Glueck and
Glueck, 1950; Cloward and Ohlen, 1960; Fleisher, 1966; Hirschi, 1969; Radsinowski and Wolfgang, 1971; Hartjen, 1974).
Silberman (1978) sums up the argument beautifully: "the high
level of criminal violence among members of the lower class
stems from poor black and Hispanic youth's sense of impotence

and exclusion. . . . If criminal violence is to be reduced to a tolerable level, those who now feel excluded must become full participating members of the American society with a major stake in its preservation" (pp. 169–70).

Silberman (1978) remarks that in the United States the "upsurge in criminal violence that began around 1960 . . . can be explained, in part, by a new and extraordinary demographic change that occurred between 1960 and 1975: the population aged fourteen to twenty-four grew 63 percent, more than six times the increase in all other age groups combined. In 1960, fourteen- to twenty-four-year-olds accounted for 69 percent of all arrests for serious crimes, although they comprised only 15.1 percent of the population" (p. 31). These numbers would have increased violent crimes by 40 to 50 percent between 1960 and 1975, but the actual increase was more than 200 percent. He terms the "greater frequency with which members of every age group, but particularly the young, commit serious crimes" a matter of "demographic overload." The growth in the fourteen- to twenty-four-year-old group was so enormous, he believes, relative to the growth of the adult population, that the conventional means of social control broke down. He puts the problem in terms of the responsibility of adults to socialize youth: "In each generation, adults must grapple with the problem of inducting the young into the norms and values of adult society." He quotes Norman Ryder, a Princeton University demographer, to the effect that "society at large is faced perennially with an invasion of barbarians." But these are assertions. In fact, the youth of each generation must grapple with the problem of securing a position on the ladder of affluence from which they can reasonably expect to make the climb successfully with a minimum of risk and effort. In this task they necessarily confront the establishment, or what Silberman calls "adult society," firmly entrenched and variously allied to resist the onslaught and the threat to its security. It is no less true in the world of urban technology than in the aboriginal world of initiations involving genital mutilations and enforced periods alone in the bush. When the youthful population is abnormally high, competition among youth, and between youth and the establishment, is abnormally intense. By the arguments advanced here, it is not at all surprising—rather it seems entirely predictable—that a dramatic increase in the proportion of young men seeking entrance into the world of social and economic success should cause a seemingly disproportionate increase in crime rates; nor is it surprising that an additional consequence should be tendencies of youth groups to ally themselves against the organizations of established members of society. Competition between youth and the establishment thus not only intensified with increases in the proportions of youth in the population, but changed its nature (Silberman, 1978, p. 34). The coincidence of disproportionate numbers of young people and a remote and unpopular war in which young men were expected to fight surely also had a significant effect.

Perhaps the extreme example of disenfranchised youth, and its reaction to its plight, was reviewed in the German news magazine *Der Spiegel* (Anonymous, 1978). It involves the teen-aged children of foreign immigrant workers in West Germany —from relatively impoverished countries like Turkey, Greece, and Yugoslavia. To the extent that these immigrant workers and their children are denied German citizenship, they constitute a population doomed to lifetimes of failure in comparison with the offspring of German citizens; they are a class of people who know ahead of time that their lives will be spent at menial tasks. Not surprisingly, then, boys and young men who are offspring of immigrant workers have an extraordinarily high crime rate.

In a book titled *The Behavior of Law* (1976), the Yale sociologist Donald Black presents an analysis independent of that developed here, yet supporting rather strongly the viewpoint I have advocated. As his title suggests, for purposes of analysis, Black treats law as the cultural anthropologist Leslie White (e.g., *The Science of Culture*, 1949) treated culture—as a thing apart from function, motivations, psychology, and individuals. He seeks correlates of the *quantity of law* (pp. 6–8) and tries to ascertain their effects. He defines law as "governmental social control"

(p. 2), and quantifies it chiefly by "the number and scope of prohibitions, obligations, and other standards to which people are subject, and by the rate of legislation, litigation, and adjudication" (p. 3). He then examines the correlates of quantitative variations in law in different circumstances and societies, and emphasizes twenty-five or thirty such correlates, which may be condensed as follows: law is "greater" (employed more often, or more effectively) in societies and social groups that are larger, more dense, more organized, more differentiated, more complex, more stratified, and in circumstances in which there are fewer other social controls (e.g., less family control) and greater "relational" (social, genetic) distances among interactants (e.g., more law is employed during interactions between distant relatives, nonrelatives, and "strangers") than in the opposite kinds of societies, social groups, and circumstances. Within societies "more" law is directed (or law is directed more often and more effectively) at individuals and groups that are relatively low-ranking, uninfluential, transient, not "respectable," socially marginal, and more distantly related.

Black's approach treats law as a singular phenomenon whose traits can be analyzed and generalized. Because law is obviously not without function, and is not independent of the motivations of people, Black's success in locating a small number of general rules, despite the enormous variation in legal systems, suggests that a certain singularity of function, therefore of motivational background, may exist for law as a whole. That is also the argument made here. Moreover, the particular correlates discovered by Black sometimes are the same as those I have emphasized, and his findings seem to support the arguments advanced here about the origins and functions of law. Black's analysis and mine differ, curiously, in that he scarcely mentions male-female differences in the "behavior" of law, while I see it as a prominent and fundamental variant.

> *A society in which the questionings of justice cease to be a constant prod and perplexity would not be human in any sense that matters.*
> —Julius Stone, 1965, p. 355

Changes in Rules with Development of Nation-States

Earlier I argued that nepotistic behavior toward nondescendant relatives evolves out of parent-offspring interactions, and that reciprocity derives from nepotism. Now I suggest that authority, in regard to regulation of social interactions, originates in parental authority, and that parental control of resources is a major aspect of rule construction and enforcement. One can extend this argument to nepotism in the forms of assisting relatives in efforts to obtain repayment of debts owed them, avenging wrongs done to relatives, and accepting responsibility for debts incurred by relatives.

The development of nation-states correlates with (a) suppression of *extensive* nepotism and polygyny as methods of reproductive competition; (b) systems of reciprocity as the prominent social cement (both direct and indirect—Flannery's "redistributive economy," figure 12); (c) a rise in concern with law and order; (d) a rise in belief in the authority of divine beings or rules; and (e) an acceptance of nobility and divinity in leaders and rulers. I suggest that a connection exists between these changes and the original parental authority, and that the origins of divinity lie in reverence toward deceased powerful ancestors and the effort by would-be leaders to use the presumed wishes and authority of such ancestors to promote order. Such efforts would depend on a person's success at convincing others of his special knowledge of such ancestors' wishes or his ability to communicate with them. With leadership and power go, I suppose, unusual reproductive opportunities. I do not regard it as an accident that God should have come to be regarded as a "Father in Heaven."

I have already argued that the rise of nation-states occurred as a result of the interactions of neighboring competitive and hostile groups as they expanded their alliances and cemented unities in a balance-of-power race. Now I suggest that rather than the rise of the nation-state being understandable from knowledge of its internal workings (Flannery, 1972), the internal workings of the nation-state are understandable only in terms of the reasons for its appearance—namely intergroup

TYPE OF SOCIETY	SOME INSTITUTIONS, IN ORDER OF APPEARANCE	ETHNOGRAPHIC EXAMPLES	ARCHAEOLOGICAL EXAMPLES
STATE	Local Group Autonomy · Egalitarian Status · Ephemeral Leadership · Ad Hoc Ritual · Reciprocal Economy · Unranked Descent Groups · Pan-Tribal Sodalities · Calendric Ritual · Ranked Descent Groups · Redistributive Economy · Hereditary Leadership · Elite Endogamy · Full Time Craft Specialization · Stratification · Kingship · Codified Law · Bureaucracy · Military Draft · Taxation	FRANCE · ENGLAND · INDIA · U.S.A.	Classic Mesoamerica · Sumer · Shang China · Imperial Rome
CHIEFDOM		TONGA · HAWAII · KWAKIUTL · NOOTKA · NATCHEZ	Gulf Coast Olmec Mexico (1000 B.C.) · Samarran of Near East (5300 B.C.) · Mississippian of North America (1200 A.D.)
TRIBE		NEW GUINEA HIGHLANDERS · SOUTHWEST PUEBLOS · SIOUX	Early Formative of Inland Mexico (1500–1000 B.C.) · Pre-Pottery Neolithic of Near East (8000–6000 B.C.)
BAND		KALAHARI BUSHMEN · AUSTRALIAN ABORIGINES · ESKIMO · SHOSHONE	Paleo-Indian and Early Archaic of U.S. and Mexico (10,000–6000 B.C.) · Late Paleolithic of Near East (10,000 B.C.)

Figure 12. Diagram from Flannery (1972) comparing extant and extinct societies as representative of four postulated types of society, and showing the attributes of each of the four types. The implication is that the larger, more complex societies are derived from smaller, simpler forms which at least resembled those described in the diagram.

aggression and competition. I will now examine briefly the sources and kinds of rules in the different kinds of societies compared by Flannery (1972) and Service (1971, 1975), and widely regarded as representing stages preceding the nation-state. They are bands, tribes, chiefdoms, and stratified societies (figure 12). I will draw heavily from Flannery's review, which I find consistent with other writings on this topic. In my earlier attempt at this analysis (Alexander, 1978d), my ignorance of the literature caused me to overlook a parallel analysis by Hoebel (1954, pp. 288–333) of what he called "the trend of the law." Here I shall attempt to integrate and compare Hoebel's conclusions with my own.

Bands

Flannery suggests that the only "segments" of bands are "families or groups of related families" and that their "means of integration are usually limited to familial bonds of kinship and marriage, plus common residence. Leadership is informal and ephemeral; division of labor is along the lines of age and sex; and concepts of territoriality, descent, or lineage are weakly developed." Murdock (1949) describes the smallest of such groups as two- or three-family bands among Eskimos.

In extant band societies there is little heritable wealth. Social interactions are said to be based largely on "reciprocity," but the term has been used by anthropologists who did not distinguish nepotism and reciprocity. Wiessner (1977) provides evidence that among Kalahari Bushmen, "reciprocity" is essentially limited to known genetic relatives, and is practiced more with close relatives than with distant ones. What authority there is in band societies seems to derive largely from parents and collections of parents—especially older men. Relatives defend and avenge one another; and they are expected to do so. The social cement of band societies is clearly nepotism. As Hoebel (1954) says, "As for law, simple societies need little of it. . . . Because relations are more direct and intimate, the primary, informal mechanisms of social control are more generally effective." Hoebel also notes that in the simplest societies "the simple community . . . ordinarily consists of a few closely related families who comprise a kindred. Relationship is bilateral; i.e., kinship to the mother's relatives is felt to be equally as strong as to the father's" (p. 293—see also figure 13 and Murdock, 1949).

Tribes
Tribes are larger groups

> whose segments are groups of families related by common descent or by membership in a variety of kinship-based groups (clans, lineages, descent lines, kindreds, etc). . . . Ancestors are often revered, and it is believed that they continue to take part in the activities of the lineage even after death. . . . Since "tribes," like bands, have weak and ephemeral leadership, they are further integrated (and even, it has been argued, regulate their environmental and interpersonal relations) by elaborate ceremonies and rituals. Some of these are conducted by formal "sodalities" or "fraternal orders" in which members of many lineages participate. . . . "Tribes" frequently have ceremonies which are regularly scheduled. . . . [and] may help to maintain undegraded environments, limit intergroup raiding, adjust man-land ratios, facilitate trade, redistribute natural resources, and "level" any differences in wealth which threaten society's egalitarian structure. . . . [Flannery, 1972, pp. 401–2]

It seems to me that Flannery may be describing the rudiments of laws that hold together groups of not-so-close relatives by imposing and maintaining restrictions on reproductive competition. He describes from archaeological finds "pottery masks . . . countless figurines of dancers . . . incredible accumulations of shell rattles, deer scapula rasps, turtle shell drums, conch shell trumpets . . . ," which suggest not only ceremony but the possibility of significant differentials in heritable wealth.

Chagnon (1968) has noted that Yanomamö Indians may not mention the dead. Yet the Yanomamö, he assures me in personal communication, otherwise fit Flannery's usage of tribes;

Mildred Dickemann (personal communication) has found that the same is true in certain New Guinea people. This sensitivity and apparent inconsistency regarding the use of ancestors within the (disputed) category of "tribe" may indicate the difference between allowing and not allowing succession to power and influence by identification with powerful deceased relatives. The Yanomamö tend to fission when a powerful ancestor dies, with the sizes of groups at fissioning correlated with the degree of genetic relatedness in the groups (Chagnon, 1976). Tribes discussed by Flannery, on the other hand, revere such ancestors, and, one supposes, may use them to enhance unity in their societies of groups of related families. The power of parents and the unity of nepotism thus appear still to be the major sources of authority and rules in the tribal societies to which Flannery refers. Whether dead ancestors are revered or unmentionable may actually represent a dichotomy between kinds of tribes that could help make the category of tribe more amenable to anthropological analysis.

Chiefdoms
Chiefdoms are still larger groups in which "lineages are 'ranked' with regard to each other, and men from birth are of 'chiefly' or 'commoner descent.'" Such chiefs

> are not merely of noble birth, but usually divine; they have special relationships with the gods which are denied commoners and which legitimize their right to demand community support and tribute . . . the chief . . . may be a priest . . . the office of 'chief' exists apart from the man who occupies it, and on his death the office must be filled by one of equally noble descent; some chiefdoms maintained elaborate genealogies to establish this. . . .
>
> Since lineages are also property-holding units, it is not surprising to find that in some chiefdoms the best agricultural land or the best fishing localities are "owned" by the highest-ranking lineages . . . high-ranking members of chiefdoms, reinforce their status with sumptuary goods, some of which archaeologists later recover in the form of "art works" in jade, turquoise, alabaster, gold, lapis lazuli, and so on. [Flannery, 1972, p. 403]

In chiefdoms it would appear that sources of authority have become more significant than in small groups, sometimes shifting from parental authority to deceased ancestors to gods representing extensions of such deceased ancestors. One also notices that the "office" of chief has itself become a vehicle of potential reproductive success for the individual who attains it—hence, itself a sought-after position (i.e., it is no longer strictly a vehicle for nepotism to the entire subject group, as in the case of a family patriarch). This prerogative is accepted and allowed by group members—perhaps because of the value to them of competition for the position of chief, which increases the likelihood that their leader will be a capable one:

> . . . within loosely organized tribes in which the local group is autonomous, trouble involving members of different local groups frequently brews physical violence which often leads to feuding; feud marks an absence of law, for the killing is not mutually acknowledged as a privilege-right; yet it appears that every society has some set procedure for avoiding feud or bringing it to a halt; among the more organized tribes on the higher levels of economic and cultural growth feud is frequently prohibited by the action of a central authority representing the total social interest; this never happens on the lower levels of culture. [Hoebel, 1954, p. 330]

In regard to larger and increasingly complex societies Hoebel (1954) says the following:

> Homogeneity gives way to heterogeneity. Common interests shrink in relation to special interests. Face-to-face relations exist not between all the members of the society but only among a progressively smaller proportion of them. Genealogical kinship links not all the members as it did heretofore but only a progressively smaller proportion of them. Access to material goods becomes more and more indirect, with greater possibilities for uneven allocation, and the struggle among the members of a given society for access to the available goods becomes intensified. Everything moves to increase the potentialities for conflict within the society. [P. 293]

Hoebel thus describes, as well as anyone, the significance of locating the underlying reasons for the continued growth and complexity of social groups during human history—the forces which continually overcame the disruptive influences that necessarily increased in numbers and intensity as society became increasingly complex. His own discussion (pp. 310ff.) of the relationship between internal unity and external enemies in the form of neighboring groups or tribes leaves a strong implication —though he nowhere explicitly supports this interpretation— that intergroup competition and aggression (or its threat) is this force of unity. Thus, "In interpersonal disputes there is a strong proclivity to resort to force. But there is an equally strong social interest in keeping internal resort to force within bounds. 'There is always the enemy to fight.' Aggression can be readily turned outside the tribe" (p. 310). He describes cases in which deaths within a group—even accidental ones—are "compensated" by efforts to kill someone in a neighboring group. Hoebel argues that "legal" solutions to problems in the Northwest Amerindians tend to be employed in disputes between lineages, while disputes within lineages tend to involve physical contests, shaming, or feuds. In sum, "the great legal problem for the tribes of the Northwest Coast is the restoration of balance between the local group aggregates that make up the larger society. Because the local groups themselves are small kinship bodies they are able to handle their internal personal problems on other than legal bases" (p. 316).

A problem paralleling that described earlier, with respect to tabooing ancestors in tribes while revering them in chiefdoms, seems also to characterize the similarly nebulous border between chiefdoms and nations (figure 13). As Mildred Dickemann points out (1979, and personal communication), some of the cultural traits that are most suppressed in modern technological nations (polygyny, extended family nepotism, dominance of males and elders) are also perhaps most extremely developed in the so-called "minimal" or "small" states that are most similar to modern nation-states in size, and may also sometimes be their immediate antecedents in history. The key to understanding the transitions among these societal types and sizes (and thus cultural patterns in general) must surely lie in a coordinated analysis of the distribution of resources, changes in modes of subsistence and the nature of heritable goods, the defensibility and spacing of physiogeographic and other boundaries, and the manner in which all of these variables influence cooperativeness and competitiveness, the accumulation of social experiences in individuals, and divisions of labor and dominance relations between the sexes and between older and younger people.

Nation-States

Finally, Flannery (1972) writes,

> The state is a type of very strong, usually highly centralized government, with a professional ruling class, largely divorced from the bonds of kinship which characterize simpler societies. It is highly stratified and extremely diversified internally, with residential patterns often based on occupational specialization rather than blood or affinal relationship. The state attempts to maintain a monopoly of force, and is characterized by true law; almost any crime may be considered a crime against the state, in which case punishment is meted out by the state according to codified procedures, rather than being the responsibility of the offended party or his kin, as in simpler societies. While individual citizens must forego violence the state can wage war; it can also draft soldiers, levy taxes, and exact tribute. [Pp. 403ff.]

Nepotism displays a peculiarly altered condition within the nation-state, as compared to the smaller kinds of human societies in which it may represent the basic social cement. Nepotism obviously cannot be the social cement of nation-states of millions or hundreds of millions of individuals; only reciprocity can fulfill this function; and, of course, the interactions of individuals in nation-states are always organized around barter, currency, and various kinds of legal obligations and documents which ensure that debts are paid. Within the large nation-states characterized by a socially and geographically mobile population, we retain ties chiefly to the immediate family, and we tend to identify as immediate family only our parents, offspring, and siblings. Because there is a correlation between the uniquely human phenomenon of socially imposed monogamy and the nation-state, everyone in the immediate family is related to Ego by one-half; nephews, nieces, aunts, and uncles by one-fourth; and first cousins by one-eighth. More distant relatives than first cousins are generally classed just that way—as "distant relatives." We do not usually organize into clans, and when we do we are usually regarded as behaving outside the law.

As Murdock (1949, pp. 1–2) notes, "Among the majority of the peoples of the earth . . . nuclear families are combined, like atoms in a molecule, into larger aggregates. . . . [But] the nuclear family will be familiar to the [American] reader as the type of family recognized *to the exclusion of all others* by our own society" (emphasis added).

Murdock also points out that "our own [American] society is characterized by bilateral descent and by the presence of kin groups of a typically bilateral type, technically called the *kindred* but popularly known under such collective terms as 'kinfolk' or 'relatives.' " (p. 45). Later he comments:

> The most distinctive structural fact about the kindred is that, save through accident, it can never be the same for any two individuals with the exception of own siblings. For any given person, its membership ramifies out through diverse kinship connections until it is terminated at some degree of relationship—frequently with second cousins, although the limits are often drawn somewhat closer or farther away than this and may be rather indefinite. The kindreds of different persons overlap or intersect rather than coincide. . . . Since kindreds interlace and overlap, they do not and cannot form discrete or separate segments of the entire society. Neither a tribe nor a community can be subdivided into constituent kindreds. This intersecting or non-exclusive characteristic is found only with bilateral descent. Every other rule of descent produces only clearly differentiated, isolable, discrete kin groups, which never overlap with others of their kind.
>
> One result of this peculiarity is that the kindred, though it serves adequately to define the jural rights of an individual, can

rarely act as a collectivity. One kindred cannot, for example, take blood vengeance against another if the two happen to have members in common. Moreover, a kindred cannot hold land or other property, not only because it is not a group except from the point of view of a particular individual, but also because it has no continuity over time. Hence under circumstances favorable either to the communal ownership of property or to the collective responsibility of kinsmen, the kindred labors under decided handicaps in comparison to the lineage or sib. [Murdock, 1949, p. 60; see also Parsons, 1954; Schneider and Smith, 1973; Goody, 1976]

What handicaps the kindred as an effective subgroup, however, enhances the likelihood of unity at the higher level. In effect, socially imposed monogamy, bilateral inheritance and descent-tracing, and the "kindred" appear to represent the combination of behavior systems most likely to lead to the enormous unified nation-states that today dominate the world. The reduction of kin groups to kindreds, however it begins, appears to be an effective way of making a larger government both powerful and stable. Perhaps bilateral descent and monogamy, leading to nation-states, initially became the norms in agricultural societies in regions of the world with particular ecological features, and once instituted were maintained by continual balance-of-power threats among hosts of political and economic units of ever-increasing size. Murdock reached a parallel conclusion:

The development of neolocal residence [not with the family of either husband or wife], in societies following other rules, appears to be favored by any influence which tends to isolate or to emphasize the individual or the nuclear family [e.g., agriculture]. Since the nuclear family is somewhat submerged under polygyny, any factor which promotes monogamy will likewise favor neolocal residence. Examples of such factors include a sex division of labor in which the product of one woman's and that of one man's activities strike an approximately equal balance, widespread poverty which inhibits extensive wife-purchase, and the adoption of a religious or ethical system, such as Christianity which forbids polygyny on principle. Since the nuclear family is also partially submerged in the presence of extended families or clans, any influence which tends to undermine or inhibit large local aggregations of kinsmen will create conditions favorable to neolocal residence. Political evolution from a gentile to a territorial form of the state, for example, has frequently been followed, in Africa, Asia, and Europe, by the disintegration of clans and the weakening of unilinear ties. Individualism in its various manifestations, e.g., private property, individual enterprise in the economic sphere, or personal freedom in the choice of marital partners, facilitates the establishment of independent households by married couples. A similar effect may be produced by overpopulation and other factors which stimulate individual migration, or by pioneer life in the occupation of new territory, or by the expansion of trade and industry, or by developing urbanization. A modification in inheritance rules, such as the replacement of primogeniture by the division of an estate among a number of heirs, can likewise favor neolocal residence. Even a change in architecture might exert an influence, e.g., the supplanting of large communal houses by a form of dwelling suited to the occupancy of a single family. [Murdock, 1949, p. 203]

AN ASIDE ON HUMAN EVOLUTION

Monogamy, bilateral inheritance, prohibition of cousin marriages, absence of extended clans, and the kin system termed the "kindred" by anthropologists thus tend to characterize both the smallest and the largest—the simplest and in some ways the most complex of human social groups—small bands like those of Eskimos, Lapplanders, and Andaman Islanders, and all of the largest technological nation-states (figure 13). In light of the difficulty, discussed in the last chapter, of drawing up a single comparative picture of all kinds of human societies all across the world and for all history, it seems a useful question—perhaps unanswerable at this stage—whether the similarities at the two extremes of societal complexity are simply analogous or represent some kind of historical continuity. In other words, in what sense or to what extent did the ancestral societies from

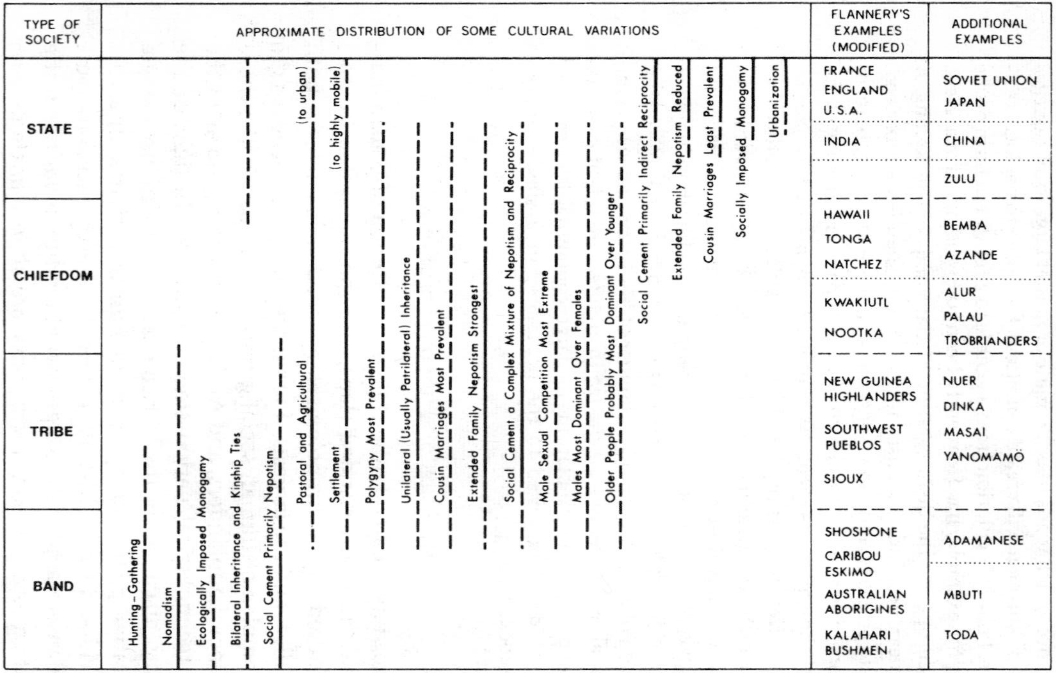

Figure 13. A diagram drawn to parallel that of Flannery (1972) (shown in figure 12), but including a different array of cultural institutions, emphasizing traits a biologist might wish to see presented. The dotted lines on the right side of the diagram are intended to emphasize the difficulty in establishing a satisfactory boundary between different kinds and sizes of socio-political units, particularly when comparing the cultural attributes portrayed in this diagram. Similarly the dashed lines indicating trait distributions represent areas of doubtfulness or greater frequencies of deviations from the suggested correlation. This diagram should be regarded as no more than a tentative hypothesis or approximate starting point.

which the modern nation-states are derived pass through stages resembling those of modern tribes and chiefdoms? Is it possible that those stages were in many cases relatively brief and not typified by the extents of clanship, cousin marriage, and asymmetry in the laterality of inheritance and lineage tracing so typical of modern societies of intermediate size in particular ecological situations in Australia, Africa, the Pacific Islands, and the New World within historical times?

This question may relate to another that is still unanswered: students of human evolution sometimes implicitly or explicitly assume that the human species and its cultural capacity and accoutrements evolved independently or in parallel in different parts of the world. Coon (1963) probably is the most explicit exponent of this view, which seems strange to biologists (and perhaps also to most social scientists) because of the exceedingly remote likelihood that the process of genetic change by mutation and selection could produce the same sets of genetic units independently in different localities. Rarely, however, have the alternatives to this biologically unlikely scenario been discussed. At another extreme the human species could have evolved in one locality and spread, replacing the existing forms (by definition in this model belonging to distinct species) without interbreeding with them. In this scenario all of the differences among modern human populations of different geographic origins would have resulted from divergences after the initial spread of the newly evolved human species.

A model somewhere between these two extremes would have each major step in the transition from a prehuman species to a human one occurring in one locality but with the subsequent spread of the new form taking place with not only extensive dispersal but much interbreeding as it spread. This hypothesis, which seems to me more parsimonious partly because of its flexibility, could include many successive waves of dispersing populations, sometimes interbreeding a great deal, sometimes little or not at all with the displaced forms. It could include a continuous genetic compatibility throughout the entire history of evolution from a prehuman or nonhuman ancestor to *Homo sapiens sapiens,* even if actual interbreeding did not always occur with contact and displacement. It could also mean that some sets of genes involved in the physical differences among human populations could have existed throughout human evolution, even though attributes like the capacity for culture and other aspects of our behavioral background were maintaining a much higher degree of uniformity across the globe. This last sugges- tion is not inconsistent with biological evolutionary theory, since it is quite possible that there has been a greater uniformity of the selective environment in regard to mental attributes than in regard to physical ones. Human mental attributes depend upon the social environment and are characterized by the de- velopment of a remarkably distinctive plasticity associated with neoteny and long juvenile life. On the other hand many physi- cal attributes, such as skin color and various aspects of body build, have obviously been selected differently in different cli- mates and do not necessarily involve great plasticity.

Definitive answers to the questions I am raising are not yet available. Perhaps the wedding of biological and anthropologi- cal approaches to long-term human history that now seems within our grasp will enable us for the first time to bring to- gether the data of physical anthropology, archaeology, and pa- leontology with that of social and cultural anthropology in a fashion that is consistent with biological evolutionary theory, and with our general knowledge and understanding of recent human history, current events, and social-ethical problems.

More on Nation-States

Returning to the nation-state it seems that in the large tech- nological forms, at least, nepotism is focused on the individual and the immediate family. In modern societies with the high and novel degree of mobility of our own, the vestiges of nepo- tism outside the family are likely to be "misdirected," in repro- ductive terms, to neighbors, roommates, or others whose social relationships to us mimic those of relatives in the past. The functional relationships between nepotism and reciprocity described earlier thus correspond to their roles in the various human social systems extant today and are also believed to correspond roughly to stages in the development of nation-states. Stein and Shand (1974) provide a closely parallel comment, from which the following is excerpted:

> As societies develop into the nation-states, they cease to be collections of fellowship groups. These groups are replaced by less personal types of social relationships, in which the members feel no special regard for each other. In the newer relationships respect for persons cannot be taken for granted. Circumstances require that people be treated as individuals, and the position of the individual in society must be recognised by the law. Further, the precise character of the law is best adapted to a society whose members are treated as separate individuals rather than as members of groups. Historically, as laws have become more sophisticated, the more they have tended to make the individual rather than the group the focus of rights and duties. These considerations do not, however, imply the attribution of a particular value to the individual as against society. [Pp. 114–16]

The existence of guarantees of reciprocity in the form of laws and written or other kinds of long-lasting undeniable contracts, as in modern nation-states, affects the relationship between reciprocity and nepotism. Without such guarantees reciprocity is more likely to be profitable with a relative than with a non-relative, unless other variables are asymmetrical; if one does not receive payment in resources he may be at least partly compensated by the genetic reproduction of the relative to whom he has extended beneficence. This fact implies that in societies without inviolable, long-lasting guarantees most social interactions should be between relatives. With enforceable guarantees nepotism may sometimes become more risky than transactions formally calling for reciprocity, because relatives may successfully prevail upon one's kindnesses outside the realm of legal guarantees; in other words, they may manipulate altruistic relatives to realize their own particular interests—for example, by calling upon mutual relatives to support their claim. The risk in such nepotism increases as the interests of a beneficiary depart from those of the altruist, hence often correlate with degree of relatedness. The effect may be to press nepotism back toward the limits of the nuclear family. Hence, perhaps, the previously mentioned adage "Never do business with a relative" may imply that if one has the means to invest formally, he should do it according to law and with a nonrelative so as unequivocally to retain the right to make cost-benefit decisions in his own interests.

It seems to me that the categories into which the laws of nation-states can be arranged are commensurate with the biological arguments made above: (a) those which prevent individuals or groups from interfering too severely with the reproductive success of others, (b) those which prevent individuals or groups from dramatically enhancing their own reproductive success, and (c) those which promote industry and creativity in individuals and groups in ways that may be used, exploited, or plagiarized by the larger collective. Examples of laws in these three categories, respectively, are those concerning: (a) murder, assault, rape, kidnapping, treason, theft, extortion, breach of contract; (b) polygamy, nepotism, tax evasion, draft evasion, monopolies; and (c) patents, copyrights, and wills.

A Note on Law and Magic

At this point a word may be said on the topic of magic and its relationship to law. Hoebel (1954) discusses in a significant way the conclusions of Gillin (1934) from his study of Barama River Caribs. Hoebel first remarks that some punishments, such as the killing of an adulterer by the cuckold, "assisted by his brothers," is a "privilege-right and no retaliation is supposed to follow." He continues by noting that a man who kills or poisons another who has violated a taboo (that is, an act regarded by the members of the society as forbidden) may be regarded as doing this "by accident," and to be "merely acting as unconscious agent of the spiritual powers who constitute the effective

agent and cannot be punished. Personal antagonism is eliminated in such a case, the solidity of the group is unaffected, and
retaliation has no function" (Gillin, 1934, p. 337).

In other words, Gillin's account is a specific implication that
the "spiritual powers" actually represent the *interests of the group
as a whole:* "public opinion supports the use of poison and sorcery as means of retaliating for an unprovoked offense," and "if
a Carib who has committed an offense discovers his victim
using either poison or sorcery against him as a retaliation, he
devotes his efforts to self-protection rather than to counter
retaliation" (Gillin, 1934, p. 335). Hoebel interprets self-protection as the neutralizing of magic. Accounts of the use of
magic or sorcery in other societies seem to me to support this
general interpretation, which is not discredited by the fact that
powerful individuals (shamans, sorcerers) may successfully dis-
tort this general significance of magic to achieve their own ends
—either temporarily or in certain circumstances. The implication I wish to emphasize is that the concept of spirits or gods
may develop or be elaborated in the context of using parental
or group power to enforce unity. In this fashion the concept of
a single god, associated with absoluteness in concepts of right
and wrong, can become an extraordinarily potent vehicle of
group unity. Whenever this view succeeds, it is obviously in the
interests of group members to prevent the misuse of these same
concepts by powerful individuals to serve their own interests
when these differ from those of the group as a whole; one way
to prevent such misuse is to restrict the possible extent of power
differentials among group members, and to restrict the extent
of allowable self-interests.

I cannot resist commenting here that the ten commandments
resemble a legal prescription for the maintenance of a nation-
state. I find it easy to interpret the first four, all of which deal
with paying homage to God and not breaking his laws, as
referring to the importance of preserving the larger group. I am
impressed that 40 percent of the rules seem concerned with this
issue.

The fifth says that we should also honor our parents. This is compatible with arguments advanced so far; half of the commandments thus deal with respecting sources of authority and not tampering with current distributions of resource control. These are the commandments that include threats of retribution. Even the fifth concludes its admonition with the phrase "that thy days may be long in the land which the Lord thy God giveth thee"—probably, however, referring to the family's survival rather than that of the individual—hence, effectively referring to genes rather than individuals.

The next four commandments tell us not to kill, commit adultery, steal, or lie. The tenth tells us not even to think about it. I am particularly impressed with the tenth commandment because, in my experience, humans tend to regard first as novel and bizarre then as ludicrous and outrageous the suggestion that their evolutionary history may have primed them to be wholly concerned with genetic reproduction. How does it happen that, in the course of evolving consciousness as a state through which some of our behaviors are expressed, we are so emphatic (and public) about rejecting this seemingly ever-so-reasonable proposition? Simultaneously we seem to reject the possibility that what we are truly about could be something we had not personally considered. It makes one wonder, quite seriously, if there might not be something incompatible about telling young children all about natural selection and rearing them to be properly and effectively social in the ways that we always have.

> . . . until force and the threat of force in international relations are brought under social control by the world community, by and for the world society, they remain the instruments of social anarchy and not the sanctions of world law.
>
> We are once again, in the long history of mankind, confronted with the stark bare-bones of the law-job. It is not this time a matter of society dribbling apart if the job fails to get done. It is the prospect of ghastly explosion.
>
> —Hoebel, 1954, p. 331

SCIENCE AS A SOCIAL ENTERPRISE

If my arguments to this point are acceptable, then science may be regarded as a particular kind of activity of individuals, sometimes operating in groups, with certain unique characteristics and consequences. Its central attribute is its unusual degree of self-correction, induced by the criterion of repeatability of results. This aspect of scientific method supposedly induces the practitioners of science to explain fully the methods by which they make their discoveries and reach their conclusions. The resulting tendency for scientific findings continually to approach correctness in explanation gives an illusion that scientists are devoted to a search for truth, hence, are somehow unusually humble and altruistic. Instead, the system of investigation called science, however it may have begun, *forces* its practitioners to report their methods as well as their results, or risk being exposed as unscientific and drummed out of their profession (see also Hull, 1978). Scientists compete by striving to acquire authorship for as many of the best ideas as possible. This competition includes identifying and publishing the errors of others. As nearly all scientists are aware, the slightest taint of deliberate falsification of results or plagiarism is often enough to damage a career permanently, and may be vastly more significant than mere incompetence. I speculate that science, as a method of finding out about the universe, began as a consequence of competition among the ancients to prove their ability to comprehend cause and effect, and to meet the challenges of one another for preeminence in this enterprise and the prestige and leadership that went with it. The requirement of repeatability is what distinguishes science, indeed, diametrically opposes it to dominance or prestige by virtue of claims of divine revelations, or knowledge conferred by deities—although the two kinds of effort may exist for exactly the same reasons.

To understand why the public tolerates and supports scientists—even, sometimes, regarding science as the most prestigious of all enterprises—we must turn to the achievements of scientific investigation. These results are represented not only by all of the products of technology but by innumerable changes of attitude toward ourselves and our environment as a result of new knowledge. In some sense, essentially all of the reasons for societal affluence, and many of the reasons for our ability to achieve a modicum of serenity in the face of the uncertainties, complexities, and competitiveness engendered by the reasons for affluence, are seen as products of science. So, I suggest, science is supported for the same reason that inventors are protected by copyright and patent laws: we evidently believe, individually and collectively (or we behave as though we believe), that the discoveries made by scientists are likely to benefit all of us sufficiently to make their support worthwhile. This view of science also contributes to the impression that scientists are humble truth-seekers, who are not trying to maximize personal gain. The truth, however, is something else, as is suggested by the enormous scale on which scientists are employed directly by organizations that exist for the sole purpose of making profits.

So long as scientific discoveries represent solutions to problems that face all humans (e.g., premature death from disease), the relationship of science to any system of ethics regarded as functional and acceptable at the group level (that is, as helping everyone about equally) is clearly a harmonious one. Even a science practiced by individuals who are totally selfish in their motivations would tend to help the group involved, except when a discovery gave a scientist such personal power as to allow him to seek his own ends in conflict with those of the majority, or the group as a whole; or to the degree that scientists themselves form subgroups with common interests that differ from those of others.

Scientists employed by subgroups, such as corporations, seeking their own profit, are somewhat removed from the continual scrutiny and approval of the collective of individuals called the public. Given the view of science I have just presented, such scientists may be expected to develop and pursue lines of investigation that do not represent the interests of the group as a whole, or even of the majority of individuals within

it. Technological and other products of science that create serious problems for society, I suggest, may frequently be expected
to come from these kinds of scientific enterprises. Accordingly,
in this particular realm many problems in the relationship of
science and ethics may be expected to occur. For example, what
is the net value to society as a whole of new herbicides, insecti-
cides, patent medicines, cosmetics, or particular trends in automobiles, farm and industrial machinery, computers, appli-
ances, and office equipment? Trends in such products may
frequently proceed in directions catering to individual needs,
desires, whims, and weaknesses, such as susceptibility to novelty, desire to prolong the phenotype at whatever cost (even, in
the eyes of relatives, using all of the resources one has saved
during a lifetime), or desire to reserve, at great cost, the opportunity to reproduce far into the future (for example, through
sperm banks). Given such propensities, and the readiness of
people to accept placebos, some of the directions taken by cor-
poration-dominated science are bound to be detrimental not
only to most of the populace but to all users, while nevertheless
profitable to their creators and manufacturers, and to the stock-
holders.

These assertions, of course, do not speak to the question of
what proportion of the scientific discoveries useful to all members of society are also likely to come from scientists employed
by profit-seeking subgroups because of the profit incentive.
Also, although government scientists who create weaponry
raising the most serious of all ethical questions may seem to be
excluded, in this context they may also be regarded as employed by subgroups, since such weaponry is presumably developed explicitly for employment against the members of
other nations when leaders regard the interests of the different
nations as sufficiently in conflict.

The above view of science is entirely compatible with the
general theory of culture and sociality described earlier. It does
not appear to me to be counter-intuitive, though it is surely not
the most widely held view of science. I believe that it tends to
resolve certain paradoxes in generally held views of science.

The next question is: how does the new view of human sociality affect our understanding of ethics, and, in turn, what does the view of ethics so generated mean for the relationship of science and ethics?

The Biological Basis of Ethics

I hypothesize that ethical questions, and the study of morality or concepts of justice and right and wrong, derive solely from the existence of conflicts of interest. In social terms there are three categories of such conflict: (a) individual versus individual, (b) group versus group, and (c) group versus individual. In biological terms two kinds of returns are involved in judging conflicts of interest: (a) those coming to Ego's phenotype and (b) those coming to Ego's genotype, through the reproductive success of various relatives including offspring. In evolutionary terms, all returns are of the second kind, and, as theories of senescence and reproductive effort indicate (Williams, 1957, 1966; Hamilton, 1966), our efforts to garner the first kind of returns are expected to be shaped so as to maximize the second kind; there is no other reason for lifetimes having evolved to be finite.

The recent exacerbation of ethical questions has been caused by an accelerating tendency for discoveries from science to cause new kinds of conflict and to cause conflict in new contexts. This situation has caused us to reexamine the basis for ethical norms, seeking generalizations that may assist us in solving the new problems. The effort is actually urgent, since cultural evolution continues to accelerate in relation to organic evolution, so that we may be assured that new ethical questions will be generated at ever-increasing rates in the future.

The two major contributions that evolutionary biology may be able to make to this problem are, first, to justify and promote the conscious realization that conflicts of interest concentrated at the individual level are what lead to ethical questions, and, second, to help identify the nature and intensity of the conflicts of interest in specific cases. Undoubtedly the most dramatic and unnerving aspect of these contributions is the argument, or realization, that all conflicts of interest among individuals resolve into conflicts over the differential reproduction of genetic units, hence, that conflicts of interest exist solely because of genetic differences among individuals, or histories of genetic differences among individuals interacting in particular fashions. I emphasize that the major barrier to acceptance of this argument—absence of theories about proximate (physiological and ontogenetic) mechanisms acceptable in light of learning theory and the modifiability of human behavior—has been at least partly eliminated (see chapter 2).

The above arguments indicate that analyses of ethics, either from a descriptive approach or as an interpretation of the sources of normative ethics in the past, must be phrased from the individual's viewpoint and must bear on the problem of how the individual is most likely to maximize his inclusive fitness. This is true even if most concepts of right and wrong, most laws, norms, traditions, and reasons for courses of action, were established in generations past and are resistant to change. The inertia of culture does not remove the individual's historical reasons for striving; it only restricts or alters his manner of striving and the degree to which his ends are likely to be achieved.

In the individual's terms, then, a statement by a biologically knowledgeable investigator about the normative ethics *of yesterday*, applicable in any cultural situation, might come out as follows: I should (that is, I should be expected to) treat others so as to maximize my inclusive fitness. My treatment of relatives should be more altruistic than my treatment of nonrelatives (that is, it should be more likely to occur in situations in which return is unlikely). My treatment of both relatives and nonrelatives should be developed in terms of the effects of my actions on (a) the reproduction of relatives (including offspring), hence, the reproduction of my genes, (b) how I will be treated by those directly affected by my actions, (c) how my relatives will be treated by those affected by my actions, and (d) how I will be treated by those only observing my actions, who are likely either to be interacting with me subsequently or to be affected by the success or failure of my actions because of the observation, and, hence, acceptance or rejection of them by still others. This description is very similar to the major theme of a general theory proposed by Bandura (1969, p. 132) that human activities are "mainly governed by anticipated outcomes based on previous consequences that were directly encountered, vicariously experienced, or self-administered."

It is particularly perplexing that we must investigate the extent to which our behavior supports this hypothesis under the realization that if such goals do guide our behavior, they are nevertheless not consciously perceived, and if the hypothesis is correct, then, paradoxically, we are evolved to reject these goals whenever we are asked to evaluate them consciously. This does not mean that we *must* reject them, but that individuals not aware of all this are expected to behave as if these were their goals even if denying it is so, and that to convince them of their self-deception may be difficult, and will be most difficult for the precise activities about which they self-deceive, for the same reason that they self-deceive. The question is testable: do we or do we not behave as predicted, whether we think so or not, when we are unaware of the predictions? It is the same kind of question anthropologists always must ask when they undertake to analyze the structure of a culture alien to their own.

By these arguments the complexity of ethical issues derives not from their general basis but from the diversity and complexity of sources of conflict, and of the means by which these sources change. We are led to a division of normative ethics into those of the past—before development of the realization that genetic interests underlie conflicts of interest—and those of the future, following conscious understanding of such arguments. It is crucial that this distinction be recognized; otherwise, what I have said above will be misinterpreted as being naively deterministic.

Why should biologists, social scientists, philosophers, and historians find it so difficult, or distasteful, to accept the idea that a new understanding of the biological basis of our behavior and our history can have liberating and socially positive effects on our lives? I am inclined to suggest that the reasons are the same as those responsible for cultural inertia and the nature of science. Leaving aside the obvious virtue of some conservatism about novelty, I suggest that those of us who make our living in the subcultural arena of science are reluctant to accept new paradigms because, if they succeed, they represent someone else changing the rules in the middle of our game; we have learned how to use the system—in our professions of science and humanism—to meet our own ends, and we resent the suggestion that we must start all over again or in any sense be placed on the same level as beginners.

Perhaps as well, it has not for a long time been profitable for social scientists to entertain truly novel theories, partly because of the supposed relationship between new ways of viewing human activities and the potential for misusing them. Thus, someone has said that a natural scientist is remembered for his best ideas, a social scientist for his worst. Perhaps the new paradigm in evolutionary biology will be first absorbed into fields like economics (e.g., Hirshleifer, 1977), and by laymen, who are curious but lack the vested interests and other inhibitory baggage of much of academia.

Right and Wrong

Interpreting the concepts of right and wrong in terms of conflicts of interest is a difficult task. First, there is an implication of absoluteness about right and wrong which gives an illusion of group function to their invocation. This connotation is promoted by legislative bodies and law; by authority in the form of parents, organized religion, and other sources of power, influence, and leadership; by persistence of meanings across generations; and even by our use of the terms right and wrong in different contexts (e.g., the *right* or *wrong* distance, direction, number, or answer; a *right* line is a straight line; the *right* hand is the correct one; *right* now means precisely at this time; etc.).

Yet all of the arguments I have presented so far suggest that this implication of absoluteness and group function has some significance other than (or additional to) actual unanimity of opinion or equality of return to all individuals. What is this significance?

Parents begin instilling the ideas of right and wrong in their children, and this is probably the normal origin of the concepts for most individuals. Initially, at least, right and wrong are defined for children as whatever their parents say is right and wrong. What, though, are usual concepts of right and wrong in parents' views of their children's behavior? One might suppose that children are simply taught by their parents never to deceive, always to tell the truth, the whole truth, and nothing but the truth; therefore, that children are taught always to be altruistic toward others, to be certain that justice is afforded all those with whom they interact, and that their own interests are secondary to those of others or of the members of the group to which they belong.

Alas, it cannot be true. As we all know very well, any children so taught, who also obeyed their parents' teaching faithfully, could not be successful, at least in this society; whatever they gained personally would immediately be lost. They would be the rubes of society, of whom advantage would be taken at every turn.

I suggest something so different that it may at first sound pernicious: that parents actually teach their children how to "cheat" without getting caught. That is, parents teach their children what are "right" and "wrong" behaviors in the eyes of others, and what truth-telling and forthright behavior actually are, so that from this base of understanding children will know how to function successfully in a world in which some deceptions are profitable, some unforgivable, hence expensive, and some are difficult to detect, others easy. I suggest that parents are more likely to punish children for (a) cheating close relatives, (b) cheating friends with much to offer the family in a continuing reciprocal interaction, or (c) cheating in an obvious, bungling fashion, sure to be detected, than they are to punish them for simply cheating (I am using the word "cheating" here is a very general way, referring to any kind of social deception or taking of advantage). It is in the interests of the members of a society that other members who are either unrelated or distantly related tend to "follow the rules" and that they not be able to deceive and manipulate successfully. Contrarily, in one's close relatives the latter abilities are reproductively advantageous, provided they are developed so as to be directed chiefly at distant relatives and nonrelatives.

In other words I suggest that the concepts of right and wrong are instilled into children in such fashion as to guide them toward inclusive-fitness-maximizing behavior in the particular societies and groups within which they are growing up and are likely to live out their lives; that they are taught by parents accustomed to living by these rules; and that the courts and prisons are filled with individuals whose teachers failed to impart just these concepts of right and wrong.

The reason that the concepts of right and wrong assume an appearance of absoluteness and group-level uniformity of application, then, are that (a) on some issues there is virtual unanimity of opinion, especially when dire external threats exist, as during wartime, and (b) it is a common social strategy to assemble as a coalition those who agree, or who can be persuaded to behave as though they agree, and then promote the apparent agreement of the subgroup as gospel. On these accounts relatively few ethical questions actually seem to involve disagreements between individuals: in one fashion or another at least one party is likely to have made his grievance appear to be that of a group. This is relatively easy to accomplish if the presumed offender constitutes a potential threat to others not directly involved. We subscribe to laws against acts like rape, robbery, and usury not so much because strangers are victims as because we have assessed, consciously or unconsciously, the probability of similar victimization of ourselves, those on whom we depend, those from whom we expect to receive assistance or resources, or those through whom we are most likely to achieve reproductive success.

In this light one may ask about the source of the recent rise of attention to issues like child abuse, rape, and the rights of minorities, women, the elderly, and the mentally and physically handicapped. I suggest that, as individuals, we regard ourselves as more vulnerable in the modern, urban, technological, socially impersonal environment, in which we are increasingly surrounded by strangers, and in which bureaucracy, weaponry, and medical knowledge of new gadgetry and substances affecting the functioning of the human body and mind seem to place us increasingly at the mercy of others. I speculate that the recent growth of interest in the rights of even nonhuman organisms represents an extension of the same trend—an effort to preserve our own rights before they are directly threatened by singling out others whose rights *are* directly threatened and using their situation to develop the social machinery to protect ourselves.

EVOLUTION AND NORMATIVE ETHICS

Arguments given above, and the cited references, make it clear why I believe that evolution has more to say about why people do what they do than any other theory. In contrast, my answer to the question, "What does evolution have to say about normative ethics, or defining what people *ought* to be doing?" is "Nothing whatsoever."

I have two reasons for giving this answer. The first is that I regard humans as sufficiently plastic in their behavior to accomplish almost whatever they wish. Unfortunately the attitude is prevalent that to suppose an evolutionary background for behavior automatically implies a predictable future into which we are helplessly cast as a consequence of the ontogenetic determinism produced in us by the history of selective action on our genes. The feeling seems to be that all evolution has to offer is information about our inevitable route through history. No one wants to know all about his future, unless the knowledge, paradoxically, promises to help him change it; and most people doubt anyway that such knowledge is possible. I am sure these feelings give rise to one kind of anti-evolutionism.

People who think this way are missing the fact that the life histories of individual organisms and the fates of species are predictable, in evolutionary terms, only to the extent that environments and their effects are predictable. For a species whose individual members possess cognitive and reflective ability, and the power of conscious prediction and testing of predictions, even the knowledge of its evolutionary history, and the interpretation of its individual tendencies in different ontogenetic environments on account of that history, become parts of the environment that determine its future. Indeed, it seems to me that no other aspect of the human experience could possibly be so influential upon our future as a clear comprehension of the fine tuning of our personalities, individually and collectively, that has resulted from the inexorable process of differential reproduction during our history.

I am saying that what a knowledge of evolution really offers us, in terms of the future, is not a restriction of ontogenetic possibilities but an elaboration of life history or life style opportunities, and of collective potential for accomplishing whatever may be desired. It does this by telling us who we really are and how to become whatever we may want to become. Evolutionary understanding, then, more than anything, has the power to make humans sufficiently plastic to accomplish *whatever they wish.* This grandiose notion, of course, loses all its glamour if there is any doubt at all about the centrality of evolutionary theory as explanatory of human nature. From my personal viewpoint, however, to have discovered that I love my child, not because it shares my genes, but because I have associated with it in certain fashions, and to discover that I am likely to prefer my own child to an adopted one like it solely because of my reproductive history, are realizations that have simultaneously made me more likely to adopt a child, less likely to reproduce compulsively, more likely to reflect in a calm and reasonable fashion over tensions associated with sexual competition, more tolerant of others in connection with all of these enterprises, and, I believe, more likely to maintain an enjoyable existence, tolerable to others, as well as worthwhile to myself.

My second reason for denying that evolutionary understanding carries lessons about what we *ought* to be doing involves the background of such notions. Ethical structures have been developed throughout history without any extensive direct knowledge or conscious perception of the evolutionary process. If existing structures have in any sense converged upon what might have been generated in the presence of such understanding it has to be because individuals and collectives of individuals have identified rights and wrongs in terms of their ultimate effects upon reproductive success. I have already argued that they have done so, and I think it is obvious that this has usually occurred without any conscious knowledge of the relationship of reproductive success to either history or proximate rewards like sensations of pleasure or well-being.

Does this mean, however, that opportunities for individual reproductive success necessarily must lie at the heart of our considerations of normative ethics for the future? I can see no reason for such an assumption.

So we are returned to proximate rewards, which have formed the basis for all systems of normative ethics anyway, without any particular evidence of their connections to ultimate reproductive success. No one needs evolutionary theory to identify pain and pleasure in his own life, although such theory may clarify their significance to us. Moreover, anyone who rejects as a proximate reward to himself whatever may be identified as such from evolutionary considerations cannot, in my opinion, be wrong.

However proper systems of normative ethics are identified, then, knowledge of evolution almost surely can help to achieve whatever goals are established. It must be obvious that I think it can do this better than any other kind of knowledge. But evolutionary understanding has little or nothing to tell us about how to identify the goals. At most it may suggest that this question is destined to remain much more complex than we should like, that answers to it are destined to change rather than become simple and static, and that it will never be answerable for all time at any particular time.

Epilogue:
On the Limits of Human Nature

Three reviewers of the manuscript for this book were disappointed that I had not more explicitly attacked the problem of what constitutes human nature, identifying its limits and explaining the consequences. Each seemed to believe that this is what an account of the biology of human sociality is actually about, and that I had somehow avoided a responsibility. While writing the book, however, I did not have in mind the objectives they describe, and I find that I have an intuitive conservatism about the entire proposition. Reflecting on their criticism, and on this intuitive feeling, I offer the following comment:

As it concerns social behavior, human nature would seem to be represented by our learning capabilities and tendencies in different situations. The limits of human nature, then, could be identified by discovering those things that we cannot learn. But there is a paradox in this, for to understand human nature would then be to know how to change it—how to create situations that would enable or cause learning that could not previously occur. To whatever extent that is so, the limits of human nature become will-o'-the-wisps that inevitably retreat ahead of our discoveries about them. Even if this is not true in all respects, I believe that it must be true in some of the most important and practical ones. I regard it as illusory to identify social behavior far outside present human capabilities (or interests) and then suggest that one has somehow said something significant about the limits of human nature, and similarly illusory to note any current human failure in social matters and regard it as unchangeable. In this light I suggest that there is much in this book that deals appropriately with human nature and its limits, though it may not always be readily identifiable as such to those who have formed opinions alternative to that expressed here.

Beyond my anxiety, beyond this writing,
the universe waits, inexhaustible, inviting.

—Jorge Luis Borges

Bibliography

Alexander, R. D. 1967. Acoustical communication in arthropods. *Ann. Rev. Ent.* 12:495–526.

———. 1969. Apthropods. In *Animal communication,* ed. T. A. Sebeok, pp. 167–219. Bloomington: Indiana U. Press.

———. 1971. The search for an evolutionary philosophy of man. *Proceedings of the Royal Society of Victoria, Melbourne* 84:99–120.

———. 1974. The evolution of social behavior. *Ann. Rev. Ecol. Syst.* 5:325–83.

———. 1975a. The search for a general theory of behavior. *Behav. Sci.* 20:77–100.

———. 1975b. Natural selection and specialized chorusing behavior in acoustical insects. In *Insects, science, and society,* ed. D. Pimentel, pp. 35–77. New York: Academic Press.

———. 1977a. Evolution, human behavior, and determinism. *Proceedings of the biennial meeting of the Philosophy of Science Assoc. (1976)* 2:3–21.

———. 1977b. Natural selection and the analysis of human sociality. In *Changing scenes in the natural sciences: 1776–1976,* ed. C. E. Goulden, pp. 283–337. Bicentennial Symposium Monograph, Phil. Acad. Nat. Sci. Special Publ. 12.

———. 1977c. Review of *The use and abuse of biology* by Marshall Sahlins. *Amer. Anthrop.* 79:917–20.

———. 1978a. Evolution, creation, and biology teaching. *Amer. Biol. Teacher* 40:91–107.

———. 1978b. Biology, determinism, and human behavior: a response to Slobodkin. *Mich. Disc. in Anthrop.* 3:154–66.

———. 1978c. Natural selection and societal laws. In *The foundations of ethics and its relationship to science.* Vol. 3, *Morals, science, and society,* ed. T. Engelhardt and D. Callahan, pp. 138–82. Hastings-on-Hudson, N.Y.: Hastings Institute.

————. 1979a. Evolution and culture. In *Evolutionary biology and human social behavior: an anthropological perspective,* ed. N. A. Chagnon and W. G. Irons, pp. 59–78. North Scituate, Mass.: Duxbury Press.

————. 1979b. Evolution, social behavior, and ethics. In *The foundations of ethics and its relationship to science.* Vol. 4, ed. T. E. Engelhardt and D. Callahan. Hastings-on-Hudson, N.Y.: Hastings Institute.

————. 1979c. Natural selection and social exchange. In *Social exchange in developing relationships,* ed. R. L. Burgess and T. L. Huston. New York: Academic Press.

————. 1979d. Sexuality and sociality in humans and other primates. In *Human sexuality: a comparative and developmental perspective,* ed. A. Katchadourian, pp. 81–97. Berkeley: U. of California Press.

————. 1979e. Human sexuality and evolutionary models. In *Human sexuality: a comparative and developmental perspective,* ed. A. Katchadourian, pp. 107–12. Berkeley: U. of California Press.

————. Unpublished ms. Speciation, with special reference to the acoustical insects and amphibians.

Alexander, R. D., and G. Borgia. 1978. Group selection, altruism, and the levels of organization of life. *Ann. Rev. Ecol. Syst.* 9:449–74.

————. 1979. On the origin and basis of the male-female phenomenon. In *Sexual selection and reproductive competition in insects,* ed. M. F. and N. Blum, pp. 417–40. New York: Academic Press.

Alexander, R. D., and W. L. Brown. 1963. Mating behavior and the origin of insect wings. *Univ. Mich. Mus. Zool. Occas. Pap.* 628:1–19.

Alexander, R. D., J. L. Hoogland, R. D. Howard, K. M. Noonan, and P. W. Sherman. 1979. Sexual dimorphisms and breeding systems in pinnipeds, ungulates, primates and humans. In *Evolutionary biology and human social behavior: an anthropological perspective,* ed. N. A. Chagnon and W. G. Irons, pp. 402–35. North Scituate, Mass.: Duxbury Press.

Alexander, R. D., and R. D. Howard. In prep. *Behavior, ecology, and life histories: a reader.*

Alexander, R. D., and K. N. Noonan. 1979. Concealment of ovulation, parental care, and human social evolution. In *Evolutionary biology and human social behavior: an anthropological perspective,* ed. N. A. Chagnon and W. G. Irons, pp. 436–53. North Scituate, Mass.: Duxbury Press.

Alexander, R. D., and K. N. Noonan. Unpublished ms. Incest, culture, and natural selection.

Alexander, R. D., and P. W. Sherman. 1977. Local mate competition and parental investment in social insects. *Science* 196:494–500.

Alexander, R. D., and D. W. Tinkle. 1968. Review of *On aggression* by Konrad Lorenz and *The territorial imperative* by Robert Ardrey. *Bioscience* 18:245–48.

————, eds. In press. *Natural selection and social behavior: recent research and new theory.* New York: Chiron Press.

Alland, A. 1972. *The human imperative.* New York: Columbia U. Press.

Allee, W. C. 1932. *Animal life and social growth.* Baltimore: Williams & Wilkins.

————. 1951. *Cooperation among animals, with human implications.* New York: Schuman.

Anonymous. 1978. Auslaenderkinder—"ein sozialer Sprengsatz." *Der Spiegel,* no. 43. Hamburg, Germany.

Ashmole, N. P. 1963. The regulation of numbers of tropical ocean birds. *Ibis* 103b:458–73.

Averhoff, W. W., and R. H. Richardson. 1976. Multiple pheromone system controlling mating in *Drosophila melanogaster. Proceedings of the National Academy of Sciences* 73: 591–93.

Bandura, A. 1969. *Principles of behavior modification.* New York: Holt, Rinehart and Winston.

Barash, D. P. 1977. *Sociobiology and behavior.* New York: Elsevier North-Holland.

Barash, D. P., W. G. Holmes, and P. J. Greene. 1978. Exact versus probabilistic coefficients of relationships: some implications for sociobiology. *Amer. Nat.* 112:355–63.

Barkow, J. H. 1978. Culture and sociobiology. *Amer. Anthrop.* 80:5–20.

Barnes, R. A., J. E. Murray, and J. Atkinson. 1968. Data from the kidney transplant registry: survival of secondary renal transplants and an analysis of early renal function. In *Advances in transplantation: proc. First Intern. Cong. Transplant. Soc.,* ed. J. Dausset, J. Hamburger, and G. Mathe, pp. 351–57.

Barth, F. 1967. On the study of social change. *Amer. Anthrop.* 69:-661–69.

Bartholomew, G. A. 1952. Reproductive and social behavior of the northern elephant seal. *U. of California Publications in Zoology* 47:369–472.

Bateman, A. J. 1948. Intrasexual selection in Drosophila. *Heredity* 2:-349–68.

Benedict, R. 1934. *Patterns of culture.* New York: Houghton Mifflin.

Berkowitz, L, and E. Walster, eds. 1976. Equity theory: toward a general theory of social interaction. *Adv. Exp. Soc. Psychol.,* vol. 9.

Bigelow, R. S. 1969. *The dawn warric.s: man's evolution toward peace.* Boston: Little, Brown.

Biocca, E., 1971. *Yanoama: the narrative of a white girl kidnapped by Amazonian Indians.* Translated by D. Rhodes. New York: Dutton.

Birkhead, T. R. 1978. Behavioural adaptations to high density nesting in the Common Guillemot *Uria aalge. Animal Behaviour* 26:321–31.

Black, D. 1976. *The behavior of law.* New York: Academic Press.

Blau, P. 1965. *Exchange and power in social life.* New York: Wiley.

Blick, J. E. 1977. Selection for traits which lower individual reproduction. *J. Theoret. Biol.* 67:597–601.

Boas, F. 1911. *The mind of primitive man.* New York: Macmillan.

———. 1940. *Race, language, and culture.* New York: Free Press.

Boissevain, J. 1974. *Friends of friends.* New York: St. Martin's Press.

Borgia, G., and R. D. Alexander. Unpublished ms. Polygyny and patrilineal inheritance.

Boucher, D., P. Breshnahan, K. Figlio, S. Risch, and S. Schneider. 1978. Sociobiological determinism: theme with variations. *Mich. Disc. in Anthrop.* 3:169–86.

Brown, J. L. 1966. Types of group selection. *Nature* 211:870.

———. 1974. Alternate routes to sociality in jays—with a theory for the evolution of altruism and communal breeding. *Amer. Zool.* 14:-61–78.

———. 1975. *The evolution of behavior.* New York: W. W. Norton.

Brown, J. L., and G. Orians. 1970. Spacing Patterns in Mobile Animals. *Ann. Rev. Ecol. Syst.* 1:239–62.

Burch, E. S., Jr. 1975. *Eskimo kinsmen: changing family relationships in northwest Alaska.* St. Paul: West Publishing Co.

Burling, R. 1958. Garo avuncular authority and matrilateral cross-cousin marriage. *Amer. Anthrop.* 60:743–49.

Bygott, J. D. 1972. Cannibalism among wild chimpanzees. *Nature* 238:-410–11.

Campbell, B. J., B. O'Neill, and B. Tingley. 1974. *Comparative injuries to belted and unbelted drivers of sub-compact, compact, intermediate, and standard cars.* U. of North Carolina Highway Safety Research Center.

Campbell, D. T. 1965. Variation and selective retention in socio-cultural evolution. In *Social change in developing areas: a re-interpretation of evolutionary theory,* ed. H. R. Barringer, G. L. Blankston, and R. W. Mack, pp. 19–49. Cambridge, Mass.: Schenckman.

Caplan, A. L., ed. 1978. The sociobiology debate. New York: Harper and Row.

Carneiro, R. L. 1961. Slash- and burn-cultivation among the Kuikuru and its implications for cultural development in the Amazon Basin. In *The evolution of horticultural systems in the native South America: causes and consequences. A symposium,* ed. J. Wilbert. *Antropologica* (Venezuela) 2:-47–67.

———. 1970. A theory of the origin of the state. *Science* 169:733–38.

———. 1975. Slash- and burn-cultivation. In *Warfare and the evolution of the state: a reconsideration,* ed. D. Webster. *Amer. Antiquity* 40:464–70.

Carroll, V., ed. 1970. *Adoption in Eastern Oceania.* Assoc. for Social Anthropology in Oceania, Monograph no. 1. Honolulu: U. of Hawaii Press.

Cavalli-Sforza, L., and M. W. Feldman. 1973. Models for cultural inheritance, I: group mean and within group variation. *Theoret. Pop. Biol.* 4:42-55.

Chagnon, N. A. 1968. *Yanomamö: the fierce people.* New York: Holt, Rinehart, and Winston.

------. 1976. Fission in an Amazonian tribe. *The Sciences* 16:14-18.

------. 1979a. Is reproductive success equal in egalitarian societies? In *Evolutionary biology and human social behavior: an anthropological perspective,* ed. N. A. Chagnon and W. G. Irons, pp. 374-401. North Scituate, Mass.: Duxbury Press.

------. 1979b. Mate competition, favoring close kin, and village fissioning among the Yanomamö Indians. In *Evolutionary biology and human social behavior: an anthropological perspective,* ed. N. A. Chagnon and W. G. Irons, pp. 86-131. North Scituate, Mass.: Duxbury Press.

Chagnon, N. A., and W. G. Irons, eds. 1979. *Evolutionary biology and human social behavior: an anthropological perspective.* North Scituate, Mass.: Duxbury Press.

Clark, J. P., and E. P. Wenninger. 1962. Socioeconomic class and area as correlates of illegal behavior among juveniles. *Amer. Socio. Rev.* 27:826-34.

Cloak, F. T. 1975. Is a cultural ethology possible? *Human Ecol.* 3:161-82.

Cloward, R. A., and L. E. Ohlin. 1960. *Delinquency and opportunity.* New York: Free Press.

Cohen, A. K. 1955. *Delinquent boys.* Glencoe, Ill.: Free Press.

Coon, C. S. 1963. *The origin of races.* New York: Knopf.

Cooper, E. L. 1976. *Comparative immunology.* Englewood Cliffs, N.J.: Prentice-Hall.

Cortes, J. B., and F. M. Gatti. 1972. *Delinquency and crime: a bio-psychosocial approach.* New York: Seminar Press.

Cowan, D. P. 1978. Behavior, inbreeding, and parental investment in solitary eumenid wasps (Hymenoptera: Vespidae). Ph.D. dissertation, U. of Michigan.

Daly, M., and M. Wilson. 1978. *Sex, evolution, and behavior.* North Scituate, Mass.: Duxbury Press.

Darwin, C. 1859. *On the origin of species.* A facsimile of the first edition with an introduction by Ernst Mayr, published in 1967. Cambridge, Mass.: Harvard U. Press.

------. 1871. *The descent of man and selection in relation to sex.* 2 vols. New York: Appleton.

Dawkins, R. 1976. *The selfish gene.* New York: Oxford U. Press.

------. 1977. Replicator selection and the extended phenotype. *Zeitschr. für Tierpsychol.* 47:61-76.

Dickmann, M. 1979. The reproductive structure of stratified human societies: a preliminary model. In *Evolutionary biology and human social organization: an anthropological perspective,* ed. N. A. Chagnon and W. G. Irons, pp. 321–67. North Scituate, Mass.: Duxbury Press.

Durham, W. H. 1976a. Resource competition and human aggression, part I: a review of primitive war. *Quart. Rev. Biol.* 51:385–415.

———. 1976b. The adaptive significance of cultural behavior. *Human Ecol.* 4:89–121.

———. 1979. Toward a coevolutionary theory of human biology and culture. In *Evolutionary biology and human social behavior: an anthropological perspective,* ed. N. A. Chagnon and W. G. Irons, pp. 39–58. North Scituate, Mass.: Duxbury Press.

Durkheim, E. 1933. *The division of labor in society.* Glencoe, Ill.: Free Press.

———. 1938. *The rules of sociological method.* 7th ed. Glencoe, Ill.: Free Press.

Eaton, R. 1978. The evolution of trophy hunting. *Carnivore* 1:110–21.

Ember, C. R. 1978. Myths about hunter-gatherers. *Ethnology* 17:439–48.

Emerson, R. M. 1969. Operant psychology and exchange theory. In *Behavioral sociology: the experimental analysis of social process,* ed. R. L. Burgess and D. Bushell, Jr. New York: Columbia U. Press.

Fallers, L. A. 1973. *Inequality: social stratification reconsidered.* Chicago: U. of Chicago Press.

Feldman, M. W., and R. C. Lewontin. 1975. The heritability hangup. *Science* 190:1163–68.

Ferracuti, F., and S. Dinitz. 1974. Cross-cultural aspects of delinquent and criminal behavior. In *Crime and delinquency: dimensions of deviance,* ed. M. Reidel and T. P. Thornberry. New York: Praeger.

Fisher, R. A. 1930. *The genetical theory of natural selection.* 2nd ed., 1958. New York: Dover.

Flannery, K. 1972. The cultural evolution of civilizations. *Ann. Rev. Ecol. Syst.* 3:399–426.

Fleisher, B. M. 1966. *The economics of delinquency.* New York: Quadrangle Books.

Flinn, M. V. In press. Human family structure and mating-marriage systems: an evolutionary biological analysis. In *Natural selection and social behavior: new research and theory,* ed. R. D. Alexander and D. W. Tinkle.

Ford, E. G. 1971. *Ecological genetics.* 3rd ed. London: Chapman and Hall.

Fortune, R. F. 1963. *Sorcerers of Dobu.* New York: Dutton.

Fouts, R. S. 1973. Acquisition and testing of gestural signals in four young chimpanzees. *Science* 180:978–80.

Freeland, W. J. 1976. Pathogens and the evolution of primate sociality. *Biotropica* 8:12–24.

Friedmann, W. 1967. *Legal theory.* 5th ed. London: Stevens & Sons.

Gallup, G. G. 1970. Chimpanzees: self-recognition. *Science* 167:86–87.

Gardner, B. T., and R. A. Gardner. 1969. Teaching sign language to a chimpanzee. *Science* 165:664–72.

———. 1971. Two-way communication with an infant chimpanzee. In *Behavior of non-human primates,* ed. A. M. Schrier and F. Stollnitz. New York: Academic Press.

Ghiselin, M. T. 1969. *The triumph of the Darwinian method.* Berkeley: U. of California Press.

Gillin, J. P. 1934. Crime and punishment among the Barama River Carib. *Amer. Anthrop.* 36:331–44.

Glueck, S., and E. Glueck. 1950. *Unravelling juvenile delinquency.* Cambridge, Mass.: Harvard U. Press.

Goode, W. J. 1966. Marital satisfaction and instability. In *Class, status, and power,* ed. R. Bendix and S. M. Lipset, pp. 377–87. 2nd ed. London: Routledge and Kegan Paul.

Goody, J. 1976. *Production and reproduction: a comparative study of the domestic domain.* Cambridge Studies in Social Anthropology, no. 17. Cambridge: Cambridge U. Press.

Gould, S. J. 1978. Review of *On human nature* by E. O. Wilson. *Human Nature* (Oct.), pp. 20–28.

Grant, P. R. 1972. Convergent and divergent character displacement. *Linnaean Society of London Biological Journal* 4:39–68.

Greene, P. J. 1978. Promiscuity, paternity, and culture. *Amer. Ethnologist.* 5:151–59.

Griliches, Z. 1957. Hybrid corn: an exploration in the economics of technological change. *Econometrica* 25:501–22.

Haldane, J. B. S. 1932. *The causes of evolution.* London: Longmans, Green. Reprint ed., 1966. Ithaca: Cornell U. Press.

Hamblin, R. L., and J. L. L. Miller. 1977. *Behavioral theory in sociology: essays in honor of George C. Homans.* New Brunswick, N. J.: Transaction Books.

Hamilton, W. D. 1963. The evolution of altruistic behaviour. *Amer. Nat.* 97:354–56.

———. 1964. The genetical evolution of social behaviour, I, II. *J. Theoret. Biol.* 7:1–52.

———. 1966. The moulding of senescence by natural selection. *J. Theoret. Biol.* 12:12–45.

———. 1967. Extraordinary sex ratios. *Science* 156:477–88.

———. 1971. Geometry for the selfish herd. *J. Theoret. Biol.* 31:295–311.

———. 1972. Altruism and related phenomena, mainly in the social insects. *Ann. Rev. Ecol. Syst.* 3:193–323.

———. 1975. Innate social aptitudes of man: an approach from evolutionary genetics. In *Biosocial Anthropology,* ed. R. Fox, pp. 133–55. New York: Wiley.

Hamilton, W. J. 1969. Social aspects of bird orientation mechanisms.

In *Animal orientation and navigation,* ed. R. M. Storm, pp. 57–71. Proceedings of the 27th Annual Biol. Colloq. Corvallis: Oregon State U. Press.

Harris, M. 1971. *The rise of anthropological theory.* New York: T. Y. Crowell.

Harris, M., and E. O. Wilson. 1978. Encounter: the envelope and the wig. *The Sciences* 18:9–15, 27–28.

Hartjen, C. A. 1974. *Crime and criminalization.* New York: Praeger.

Hartung, J. 1976. On natural selection and the inheritance of wealth. *Current Anthrop.* 17:607–22.

Hatch, M. 1973. *Theories of man and culture.* New York: Columbia U. Press.

Hatfield, E. M., M. K. Utne, and J. Traupman. 1979. Equity. In *Social exchange in developing relationships,* ed. R. Burgess and R. Huston. New York: Academic Press.

Hill, J. L. 1974. Peromyscus: effect of early pairing on reproduction. *Science* 186:1042–44.

Himelhoch, J. 1972. A psychosocial model for the reduction of lower-class youth crime. In *Crime prevention and social control,* ed. R. L. Akers and E. Sagarin, pp. 3–14. New York: Praeger.

Hirschi, T. 1969. *Causes of delinquency.* Berkeley: U. of California Press.

Hirshleifer, J. 1977. Economics from a biological viewpoint. *J. of Law and Economics* 20:1–52.

Hoebel, E. A. 1954. *The law of primitive man: a study of comparative legal dynamics.* Cambridge, Mass.: Harvard U. Press.

Hofstadter, R. 1955. *Social Darwinism in American thought.* Boston: Beacon.

Homans, G. C. 1961. *Social behavior: its elementary forms.* Rev. ed., 1974. New York: Harcourt, Brace & World.

Homans, G. C., and D. M. Schneider. 1955. *Marriage, authority, and final causes.* Glencoe, Ill.: Free Press.

Hoogland, J. L. 1977. The evolution of coloniality in white-tailed and black-tailed prairie dogs (Sciuridae: *Cynomys leucurus* and *C. ludovicianus*). Ph. D. dissertation, U. of Michigan.

————. In press, a. Aggression, ectoparasitism, and other possible costs of prairie dog (Sciuridae: *Cynomys* spp.) coloniality. *Behaviour.*

————. In press, b. The effects of colony size on individual alertness of prairie dogs (Sciuridae: *Cynomys* spp.) *Animal Behaviour.*

————. In press, c. The evolution of coloniality in prairie dogs (Sciuridae: *Cynomys* spp.).

Hoogland, J. L., and P. Sherman. 1976. Advantages and disadvantages of bank swallow *(Riparia riparia)* coloniality. *Ecol. Monographs* 46:-33–58.

Horn, H. S. 1971. Social behavior of nesting Brewer's Blackbirds. *Condor* 72:15–23.

Howard, R. D. 1979a. Early embryo mortality in bullfrogs. *Ecology.*

——. 1979b. Evolution of mating strategies. *Evolution.*

——. In press. Estimating fitness in natural populations. *Amer. Nat.*

Hoy, R. R., J. Hahn, and R. C. Paul. 1977. Hybrid cricket auditory behavior: evidence for genetic coupling of communication. *Science* 195:82–84.

Hull, D. L. 1978. Altruism in science: a sociobiological model of cooperative behaviour among scientists. *Animal Behav.* 26:685–97.

Irons, W. G. 1979a. Investment and primary social dyads. In *Evolutionary biology and human social behavior: an anthropological perspective,* ed. N. A. Chagnon and W. G. Irons, pp. 181–213. North Scituate, Mass.: Duxbury Press.

——. 1979b. Natural selection, adaptation, and human social behavior. In *Evolutionary biology and human social behavior: an anthropological perspective,* ed. N. A. Chagnon and W. G. Irons, pp. 4–39. North Scituate, Mass.: Duxbury Press.

Jenni, D. A. 1974. Evolution of polyandry in birds. *Amer. Zool.* 14:-129–44.

Johansen, K. 1977. Reproductive medicine. In *Immunology in medicine: a comprehensive guide to clinical immunology,* ed. E. J. Holborow and W. G. Reeves, pp. 675–707. New York: Grune and Stratton.

Keith, A. 1949. *A new theory of human evolution.* New York: Philosophy Library.

Kelsen, H. 1957. *What is justice? Justice, law, and politics in the mirror of science. Collected essays.* Berkeley: U. of California Press.

Kitchen, D. W. 1974. Social behavior and ecology of the pronghorn. *Wildlife Monograph* 38:1–96.

Kroeber, A. L. 1909. Classificatory systems of relationship. *J. of Royal Anthrop. Inst. Great Britain and Ireland* 39:77–84.

Kropotkin, P. 1902. *Mutual aid: a factor in evolution.* New York: Doubleday.

Kummer, H. 1968. *Social organization of hamadryas baboons: a field study.* Chicago: U. of Chicago Press.

Kurland, J. A. 1979. Paternity, mother's brother, and human sociality. In *Evolutionary biology and human social behavior: an anthropological perspective,* ed. N. A. Chagnon and W. G. Irons, pp. 145–80. North Scituate, Mass.: Duxbury Press.

Lack, D. 1939. *Darwin's finches.* New York: Harper and Row.

——. 1954. *The natural regulation of animal numbers.* New York: Oxford U. Press.

——. 1966. *Population studies of birds.* Oxford: Clarendon Press.

——. 1968. *Ecological adaptations for breeding in birds.* London: Methuen.

Lamb, M. J. 1977. *The biology of ageing.* New York: Wiley.

Lancaster, J. B. 1979. Sex and gender in evolutionary perspective. In

Human sexuality: a comparative and developmental perspective, ed. H. Katchadourian, pp. 51–80. Berkeley: U. of California Press.

Laslett, P., and L. Wall. 1972. *Household and family in past time.* Cambridge: Cambridge U. Press.

Lawick-Goodall, J. V. 1967. Mother-offspring relationships in free-ranging chimpanzees. In *Primate ethology,* ed. D. Morris. Chicago: Aldine.

LeBoeuf, B. J. 1974. Male-male competition and reproductive success in elephant seals. *Amer. Zool.* 14:163–67.

Lee, R. B., and I. DeVore. 1968. *Man the hunter.* Chicago: Aldine.

Leigh, E. 1977. How does selection reconcile individual advantage with the good of the group? *Proceedings of Natl. Acad. Sci.* 74:4542–46.

———. 1978. Accounting for sexual reproduction. *Science* 202:1274–75.

Levi-Strauss, C. 1969. *The elementary structures of kinship.* Boston: Beacon.

Lewontin, R. C. 1970. The units of selection. *Ann. Rev. Ecol. Syst.* 1:1–18.

Lindzey, G. 1967. Some remarks concerning incest, the incest taboo, and psychoanalytic theory. *Amer. Psychol.* 22:1051–59.

Linton, C. R. 1936. *The study of man.* New York: Appleton-Century.

Low, B. S. 1976. The evolution of amphibian life histories in the desert. In *Evolution of desert biota,* ed. D. W. Goodall, pp. 149–95. Austin: U. of Texas Press.

———. 1978. Environmental uncertainty and the parental strategies of marsupials and placentals. *Amer. Nat.* 112:197–213.

———. 1979. Sexual selection and human ornamentation. In *Evolutionary biology and human social behavior: an anthropological perspective,* ed. N. A. Chagnon and W. G. Irons, pp. 462–86. North Scituate, Mass.: Duxbury Press.

Lowie, R. H. 1920. *Primitive society.* Reprint ed., 1947 and 1970. New York: Liveright.

Luria, Z. 1979. Psychosocial determinants of gender identity, role, and orientation. In *Human sexuality: a comparative and developmental perspective,* ed. A. Katchadourian pp. 163–93. Berkeley: U. of California Press.

McClintock, M. K. 1971. Menstrual synchrony and suppression. *Nature* 229:244–45.

McCord, W. M., and J. McCord. 1969. *Origins of crime: a new evaluation of the Cambridge-Somerville youth study.* Patterson Smith Reprint Series in Criminology, Law Enforcement, and Social Problems, no. 49. Montclair, N.J.: Patterson Smith.

McMillen, M. M. 1979. Differential mortality by sex in fetal and neonatal deaths. *Science* 204:89–91.

Malinowski, B. 1926. *Crime and custom in savage society.* New York: Harcourt, Brace & World.

————. 1944. *A scientific theory of culture and other essays.* Oxford: Oxford U. Press.

Marsh, R. M. 1967. *Comparative sociology: a codification of cross-societal analysis.* New York: Harcourt, Brace & World.

Mason, W. A. 1976. Environmental models and mental modes. Representational process in the great apes and man. *Amer. Psychol.* 31:-284–94.

Masters, R. D. Unpublished ms. Classical political philosophy and contemporary biology.

Mauss, M. 1954. *The gift: forms and functions of exchange in archaic societies.* London: Cohen and West.

Maynard Smith, J. 1964. Group selection and kin selection. *Nature* 201:1145–47.

————. 1976. Group selection. *Quart. Rev. Biol.* 31:277–83.

————. 1978. *The evolution of sex.* Cambridge: Cambridge U. Press.

Maynard Smith, J., and G. R. Price. 1973. The logic of animal conflict. *Nature* 246:15–18.

Mayr, E. 1975. Behavioral programs and evolutionary strategies. *Amer. Sci.* 62:650–59.

Mech, L. D. 1970. *The wolf: the ecology and behavior of an endangered species.* New York: Natural History Press.

Medawar, P. 1955. The definition and measurement of senescence. *Ciba Found. Colloq. on Ageing* 1:4–15.

————. 1957. *The uniqueness of the individual.* London: Methuen.

Merton, R. K. 1968. *Social theory and social structure.* Enlarged ed. New York: Free Press.

Miller, W. B. 1958. Lower class culture as a generating milieu of gang delinquency. *J. Social Issues* 14:5–19.

Money, J. J., E. Cawk, G. N. Niancki, and B. Nuncombe. 1970. Sex training and traditions in Arnhem Land. *Brit. J. Med. Psychol.* 47:-383–99.

Money, J. J., and A. A. Ehrhardt. 1972. *Man and woman, boy and girl.* Baltimore: John Hopkins U. Press.

Montagu, M. F. A. 1955. *The direction of human development.* New York: Harper Bros.

————. 1976. *The nature of human aggression.* New York: Oxford U. Press.

Morgan, L. H. 1871. Systems of consanguinity and affinity of the human family. *Smithsonian Contributions to Knowledge* 17:4–602.

Mulvihill, D. J., and M. M. Tumin. 1969. *Crimes of violence.* Washington, D.C.: U.S. Govt. Printing Office.

Murdock, G. P. 1949. *Social structure.* New York: Macmillan.

————. 1960. How culture changes. In *Man, culture, and society,* ed. H. L. Shapiro, pp. 247–60. New York: Oxford U. Press.

————. 1967. *Ethnographic atlas.* Pittsburgh: U. of Pittsburgh Press.

————. 1972. Anthropology's mythology. *Proceedings of Royal Anthrop. Inst. of Great Britain and Ireland for 1971*, pp. 17–24.

Muul, I. 1968. *Behavioral and physiological influences on the distribution of the flying squirrel*, Glaucomys volans. U. of Mich. Zool. Misc. Publ. 134.

Oster, G. F., and E. O. Wilson. 1979. *Caste and ecology in the social insects.* Princeton, N.J.: Princeton U. Press.

Parsons, T. 1954. *Essays in sociological theory.* Rev. ed. Glencoe, Ill.: Free Press.

Posposil, L. 1958. *Kapauku Papuans and their law.* New Haven: Yale U. Publ. in Anthrop., no. 54.

Power, H. W. 1976. On forces of selection in the evolution of mating types. *Amer. Nat.* 110:937–44.

Premack, D. 1971. On the assessment of language competence in the chimpanzee. In *Behavior of nonhuman primates,* ed. A. M. Schrier and F. Stollnitz, pp. 185–228. New York: Academic Press.

Radcliffe-Brown, A. R. 1922. *The Andaman islanders.* Reprint ed., 1964. New York: Free Press.

————. 1924. The mother's brother in South Africa. *S. Afr. J. Sci.* 21:542–55.

————. 1951. Murngin social organization. *Amer. Anthrop.* 5:37–55.

————. 1952. *Structure and function in primitive society.* Glencoe, Ill.: Free Press.

Radsinowski, L., and M. E. Wolfgang, eds. 1971. *Crime and justice: the criminal in society.* New York: Basic Books.

Rappaport, R. 1968. *Pigs for the ancestors: ritual in the ecology of a New Guinea people.* New Haven: Yale U. Press.

Rawls, J. 1971. *A theory of justice.* Cambridge, Mass.: Harvard U. Press.

Richerson, P. J., and R. Boyd. 1978. A dual inheritance model of the human evolutionary process, I: basic concepts and a simple model. *J. Human and Social Biol. Struct.* 1:127–54.

————. In press. A dual inheritance model of the human evolutionary process, II: costly culture and the genetic control of cultural fitness. *Sociobiology and Behavioral Ecology.*

Rosenfeld, A. 1977. When man becomes as God: the biological prospect. *Sat. Review* 5:15–20.

Rumbaugh, D. M., T. V. Gill, and E. C. von Glaserfeld. 1973. Reading and sentence completion by a chimpanzee *(Pan). Science* 182:731–33.

Sahlins, M. D. 1965. On the sociology of primitive exchange. In *The relevance of models for social anthropology,* ed. M. Banton, pp. 139–236. London: Tavistock.

————. 1976a. *The use and abuse of biology: an anthropological critique of sociobiology.* Ann Arbor: U. of Michigan Press.

————. 1976b. *Culture and practical reason.* Chicago: U. of Chicago Press.

Schaller, G. B. 1972. *The Serengeti lion.* Chicago: U. of Chicago Press.

Schneider, D. M. 1968. *American kinship: a cultural account.* Englewood Cliffs, N. J.: Prentice Hall.

———. 1972. What is kinship all about? In *Kinship studies in the Morgan Centennial Year,* ed. P. Reinig, pp. 32–62. Anthropological Society of Washington, D.C.

Schneider, D. M., and C. B. Cottrell. 1975. *The American kin universe: a genealogical study.* U. of Chicago Dept. of Anthrop., Studies in Anthropology, Series in Social, Cultural, and Linguistic Anthropology, vol. 3.

Schneider, D. M., and K. Gough, eds. 1961. *Matrilineal kinship.* Reprinted., 1974. Berkeley: U. of California Press.

Schneider, D. M., and R. T. Smith. 1973. *Class differences and sex roles in American kinship and family structure.* Englewood Cliffs, N.J.: Prentice Hall.

Seger, J. 1976. Evolution of responses of relative homozygosity. *Nature* 262:578–80.

Service, E. R. 1971. *Cultural evolutionism: theory in practice.* New York: Holt, Rinehart and Winston.

Service, E. R. 1975. *Origins of the state and civilization: the process of cultural evolution.* New York: W. W. Norton.

Shapiro, M. F. 1958. *Children of the kibbutz.* Cambridge, Mass.: Harvard U. Press.

Sheldon, W. H. 1961. The criterion of the good *and* right. In *Experience, existence and the good,* ed. I. C. Lieb, pp. 275–84. Carbondale: Southern Illinois U. Press.

Shepfer, J. 1971. *Self-imposed incest avoidance and exogamy in second-generation kibbutz adults.* Ann Arbor, Michigan: University Microfilms.

———. 1978. Reflections on the origin of the human pair-bond. *J. Soc. Bio. Struct.* 1:253–64

Sherman, P. W. 1977. Nepotism and the evolution of alarm calls. *Science* 197:1246–53.

———. In press. The limits of nepotism. In *Sociobiology: beyond nature-nurture,* ed. G. W. Barlow and J. Silverberg. AAAS Publ.

Siegel, S. 1956. *Nonparametric statistics for the behavioral sciences.* New York: McGraw-Hill.

Silberman, C. E. 1978. *Criminal violence, criminal justice.* New York: Random House.

Simpson, R. C. 1972. *Theories of social exchange.* Morristown, N.Y.: General Learning Press.

Sladen, W. J. L. 1955. *The pygoscelid penguins.* Sci. Rep. 17. London: F.I.D.S.

Slobodkin, L. B. 1977. Problems on the border between biological and social sciences. *Mich. Disc. in Anthrop.* 2:124–37.

Spiro, M. E. 1958. *Children of the kibbutz.* Cambridge, Mass.: Harvard U. Press.

———. 1961. Social systems, personality, and functional analysis. In *Studying personality cross-culturally,* ed. B. Kaplan, pp. 93–197. New York: Harper and Row.

Staples, R. 1978. Masculinity and race: the dual dilemma of black men. *J. of Social Issues* 34:169–83.

Stein, P., and J. Shand. 1974. Legal values in western society. Edinburgh: Edinburgh U. Press.

Stent, G. 1978. Review of *The selfish gene* by Richard Dawkins. Hastings-on-Hudson, N.Y.: Hastings Institute.

Stone, J. 1965. *Human law and human justice.* Stanford: Stanford U. Press.

Sutherland, E. H., and R. Cressey. 1966. *Principles of criminology.* 7th ed. New York: Lippincott.

Suzuki, D. T., and A. J. F. Griffiths. 1976. *An introduction to genetic analysis.* San Francisco: W. H. Freeman.

Talmon, Y. 1965. The family in a revolutionary movement—the case of the kibbutz in Israel. In *Comparative family systems,* ed. M. F. Nimkoff, pp. 259–86. Boston: Houghton Mifflin.

Tener, J. S. 1965. *Muskoxen in Canada: a biological and taxonomic review.* Ottawa: Queen's Printer.

Trivers, R. L. 1971. The evolution of reciprocal altruism. *Quart. Rev. Biol.* 46:35–57.

———. 1972. Parental investment and sexual selection. In *Sexual selection and the descent of man,* ed. B. Campbell, pp. 136–79. Chicago: Aldine.

———. 1974. *Parent-offspring conflict. Amer. Zool.* 14:249–64.

Trivers, R. L., and H. Hare. 1976. Haplodiploidy and the evolution of the social insects. *Science* 191:249–63.

Trivers, R. L., and D. E. Willard. 1973. Natural selection of parental ability to vary the sex ratio of offspring. *Science* 179:90–92.

Underhill, R. M. 1939. *Social organization of the Papago Indians.* Columbia U. Contributions to Anthropology, no. 30. Reprint ed., 1969. New York: AMS Press.

Van den Berghe, P., and D. P. Barash. 1977. Inclusive fitness and human family structure. *Amer. Anthrop.* 79:809–23.

Wade, M. S. 1976. Group selection among laboratory populations of *Tribolium. Proceedings of Natl. Acad. Sci.* 73:4604–07.

Wade, N. 1976. Sociobiology: troubled birth for new discipline. *Science* 191:1151–55.

Walker, T. J. 1974. Character displacement and acoustical insects. *Amer. Zool.* 14:1137–50.

Wallace, A. F. C. 1961. The psychic unity of human groups. In *Studying personality cross-culturally,* ed. B. Kaplan, pp. 129–63. New York: Harper and Row.

Washburn, S. L. 1978. Human behavior and the behavior of other animals. *Amer. Psychol.* 33:405–18.

Webster, D. 1975. Warfare and the evolution of the state: a reconsideration. *Amer. Antiquity* 40:464–70.

Weinrich, J. D. 1977. Human sociobiology: pair-bonding and resource predictability (effects of social class and race). *Behav. Ecol. and Sociobol.* 2:91–118.

West Eberhard, M. J. 1975. The evolution of social behavior by kin selection. *Quart. Rev. Biol.* 50:1–33.

———. 1976. Born: sociobiology. A review of *Sociobiology* by E. O. Wilson. *Quart. Rev. Biol.* 51:89–92.

Wheeler, W. M. 1923. *Social life among the insects.* New York: Harcourt, Brace.

White, L. A. 1949. *The science of culture: a study of man and civilization.* Reprinted. 1969. New York: Farrar, Straus & Giroux.

White, L. A. 1959. *The evolution of culture: the development of civilization to the fall of Rome.* New York: McGraw-Hill.

Wickler, W. 1968. *Mimicry in plants and animals.* New York: McGraw-Hill.

Wiessner, P. 1977. Hxaro: a regional system of reciprocity among the !Kung San for reducing risk. Ph.D. dissertation, U. of Michigan.

Williams, G. C. 1957. Pleiotropy, natural selection, and the evolution of senescence. *Evolution* 11:398–411.

———. 1966. *Adaptation and natural selection.* Princeton: Princeton U. Press.

Williams, G. C., and D. C. Williams. 1957. Natural selection of individually harmful social adaptations among sibs with special reference to social insects. *Evolution* 11:32–39.

Wilson, D. S. 1975a. New model for group selection. *Science* 189:8701.

———. 1975b. A theory of group selection. *Proceedings of Nat. Acad. Sci.* 72:143–46.

Wilson, E. O. 1973a. The queerness of social evolution. *Bull. Entomol. Soc. Am.* 19:20–22.

———. 1973b. Group selection and its significance for ecology. *Bioscience* 23:631–38.

———. 1975. *Sociobiology: the new synthesis.* Cambridge, Mass.: Harvard U. Press.

———. 1978. *On human nature.* Cambridge, Mass.: Harvard U. Press.

Wolf, A. P. 1966. Childhood association, sexual attraction, and the incest taboo: a Chinese case. *Amer. Anthrop.* 68:883–98.

———. 1968. Adopt a daughter-in-law, marry a sister: a Chinese solution to the problem of the incest taboo. *Amer. Anthrop.* 70:-864–74.

Wright, H. T. 1977. Recent research on the origin of the state. *Ann. Rev. Anthrop.* 6:379–97.

Wynne-Edwards, V. C. 1962. *Animal dispersion in relation to social behaviour.* Edinburgh: Oliver & Boyd.

Index of Names

Index of Subjects